微信小程序开发

图解案例教程

附精讲视频 | 第2版

刘刚 | 著

人民邮电出版社

北京

图书在版编目（CIP）数据

微信小程序开发图解案例教程：附精讲视频／刘刚著．－－2版．－－北京：人民邮电出版社，2019.1
ISBN 978-7-115-48987-6

Ⅰ．①微… Ⅱ．①刘… Ⅲ．①移动终端－应用程序－程序设计－教材 Ⅳ．①TN929.53

中国版本图书馆CIP数据核字(2018)第175760号

内 容 提 要

本书分两篇，介绍了微信小程序设计的基本知识和实战案例。第一篇为微信小程序快速入门，包括认识微信小程序、微信小程序框架分析、用微信小程序组件构建 UI 界面、必备的微信小程序 API、微信小程序设计及问答；第二篇为综合案例应用，包括仿智行火车票 12306 微信小程序、仿糗事百科微信小程序、仿中国婚博会微信小程序 3 个综合实战案例。本书采用图、表与详细说明的示例代码相结合的叙述方式，讲解微信小程序设计的基本原理和知识，简单易懂，提供了丰富详尽的实战案例，带读者边做边学，快速掌握微信小程序的设计和实现。

本书可供对微信小程序开发有兴趣的读者自学，也可作为院校、培训机构关于微信小程序开发课程的教材。

◆ 著 刘 刚
 责任编辑 桑 珊
 责任印制 马振武

◆ 人民邮电出版社出版发行　北京市丰台区成寿寺路 11 号
 邮编 100164　电子邮件 315@ptpress.com.cn
 网址 http://www.ptpress.com.cn
 山东百润本色印刷有限公司印刷

◆ 开本：787×1092 1/16
 印张：22.75　　　　　　　2019 年 1 月第 2 版
 字数：527 千字　　　　　　2019 年 1 月山东第 1 次印刷

定价：69.80 元

读者服务热线：(010)81055256　印装质量热线：(010)81055316
反盗版热线：(010)81055315
广告经营许可证：京东工商广登字 20170147 号

第2版前言
Foreword

为什么要学微信小程序

微信小程序是微信团队在2017年1月9日正式发布的功能，它可以实现App软件的原生交互操作效果，但是不像App软件需要下载安装才能使用。微信小程序只需要用户扫一扫或者搜一下就可以使用，不仅符合用户的使用习惯，还释放了用户手机的内存空间，同时给企业提供了宣传自己产品的渠道。创建微信小程序就可以被更多用户找到自己的产品，宣传自己的产品。微信小程序的快速发展，为我们提供了很多的就业机会。让我们赶快成为一名小程序员吧！

使用本书，你可以通过3步学会微信小程序

Step1 通过图、文、代码、视频精讲快速理解小程序的基本原理和应用方法。

学什么，用来做什么 →

3.6 地图组件

map地图组件用来开发与地图有关的应用，如地图导航、打车软件、京东商城的订单轨迹都会用到地图组件，在地图上可以标记覆盖物以及指定一系列的坐标位置。京东的仓库和客户的收货地址，如图3.58所示。

扫码看精讲视频

地图组件

图3.58 京东订单轨迹

表3.36 map地图属性

属性	类型	默认值	说明
longitude	Number		中心经度
latitude	Number		中心纬度
scale	Number	16	缩放级别，取值范围为5~18
markers	Array		标记点
covers	Array		即将移除，请使用 markers
autoplay	Boolean		是否自动播放
polyline	Array		路线
circles	Array		圆
controls	Array		控件
include-points	Array		缩放视野以包含所有给定的坐标点
show-location	Boolean		显示带有方向的当前定位点
bindmarkertap	EventHandle		单击标记点时触发
bindcontroltap	EventHandle		单击控件时触发
bindregionchange	EventHandle		视野发生变化时触发
bindtap	EventHandle		单击地图时触发

→ 图+表详解基本原理

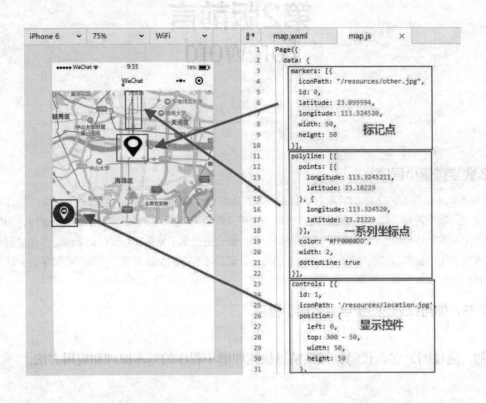

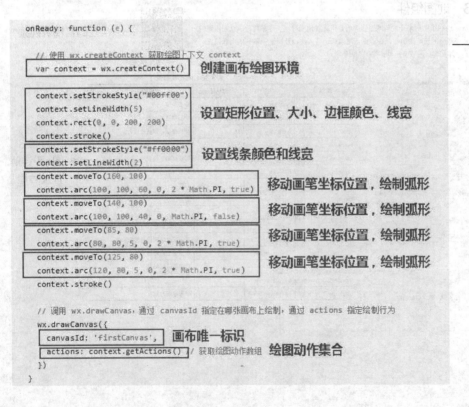

案例代码详细说明，一看就懂

Step2 沙场大练兵——边做边学。

4.13 沙场大练兵：仿豆瓣电影微信小程序

豆瓣电影App是一款用来购买电影票、查看影评的软件，图4.33、图4.34、图4.35、图4.36是豆瓣电影主要界面。

下面我们来设计一款豆瓣电影微信小程序，可以查看上映的电影以及电影详情内容。

仿豆瓣电影微信小程序

——○ 学完马上实战演练

图4.33 电影　　　　图4.34 影院

——○ 案例最终效果，学完本章就会做

Step3 综合实战，感受真实商业项目制作过程。

第二篇　综合案例应用

Chapter06 第6章

综合案例：仿智行火车票12306微信小程序

——○ 完整运用所学知识

智行火车票是一款收取费用的自动查询预订火车票的软件，可以实时监控票数的多少，与铁路部门数据实时同步，该软件可以直接在12306上订票，可以查询、预定和购买。本章案例模仿制作智行火车票小程序。

- 需求描述
- 设计思路及相关知识点
- 准备工作
- 设计流程
- 小结

商业项目零
距离接触

不只讲实现，
还讲调研和
设计

图6.1 火车票　　　　图6.2 飞机票

小刚（刘刚）老师简介

- 一线项目研发、设计、管理工程师、高级项目管理师、项目监理师，负责纪检监察廉政监督监管平台、国家邮政局项目、政务大数据等多个国家级项目的设计与开发。
- 极客学院、北风网金牌讲师。
- 畅销书《微信小程序开发图解案例教程（附精讲视频）》《小程序实战视频课：微信小程序开发全案精讲》《Axure RP8原型设计图解微课视频教程（Web+App）》作者。

平台支撑，免费赠送资源

- 全部案例源代码、全书电子教案可登录人邮教育社区（www.ryjiaoyu.com.cn）或网盘（https://box.lenovo.com/l/q5WSJC，提取码：028a）下载。
- 全书高清精讲视频课程（扫描书中二维码或登录人邮学院观看，人邮学院登录方法见本书封底）。

著者
2018年10月

目录
Contents

第一篇 微信小程序快速入门

第1章 认识微信小程序 1

- 1.1 微信小程序介绍 2
  - 1.1.1 初识微信小程序 2
  - 1.1.2 微信小程序的功能 3
  - 1.1.3 微信小程序的使用场景 3
  - 1.1.4 微信小程序能取代App吗 5
  - 1.1.5 微信小程序的发展历程 5
  - 1.1.6 微信小程序带来的机会 6
- 1.2 微信小程序开发准备 6
  - 1.2.1 基础技术准备 6
  - 1.2.2 开发准备 6
- 1.3 微信小程序开发工具的使用 8
  - 1.3.1 创建项目 8
  - 1.3.2 开发者工具界面 10
  - 1.3.3 模拟器区域 10
  - 1.3.4 编辑器区域 11
  - 1.3.5 调试器区域 13
  - 1.3.6 工具栏区域 15
  - 1.3.7 常用快捷键 17
- 1.4 沙场大练兵：Hello World的创建 18
- 1.5 小结 20

第2章 微信小程序框架分析 21

- 2.1 微信小程序目录结构介绍 22
  - 2.1.1 框架全局文件 22
  - 2.1.2 工具类文件 26
  - 2.1.3 框架页面文件 27
  - 2.1.4 小试牛刀：制作猫眼电影底部标签导航 28
- 2.2 微信小程序注册程序应用 29
- 2.3 微信小程序注册页面的使用 31
  - 2.3.1 页面初始化数据 32
  - 2.3.2 生命周期函数 32
  - 2.3.3 页面相关事件处理函数 32
  - 2.3.4 页面路由管理 33
  - 2.3.5 自定义函数 34
  - 2.3.6 setData设值函数 34
- 2.4 微信小程序如何绑定数据 35
  - 2.4.1 组件属性绑定 35
  - 2.4.2 控制属性绑定 35
  - 2.4.3 关键字绑定 36
  - 2.4.4 运算 36
  - 2.4.5 小试牛刀：天气微信小程序 36
- 2.5 微信小程序条件渲染 39
  - 2.5.1 wx:if判断单个组件 39
  - 2.5.2 block wx:if 判断多个组件 40
- 2.6 微信小程序列表渲染 40
  - 2.6.1 wx:for 列表渲染单个组件 40
  - 2.6.2 block wx:for 列表渲染多个组件 40
  - 2.6.3 wx:key 指定唯一标识符 40
- 2.7 微信小程序定义模板 41
  - 2.7.1 定义模板 41
  - 2.7.2 使用模板 41
- 2.8 微信小程序的引用功能 42
  - 2.8.1 import引用 42
  - 2.8.2 include引用 42
- 2.9 WXS小程序脚本语言 42
  - 2.9.1 模块化 43
  - 2.9.2 变量与数据类型 44
  - 2.9.3 注释 46
  - 2.9.4 语句 46
- 2.10 沙场大练兵：仿香哈菜谱微信小程序 48
  - 2.10.1 底部标签导航设计 49

2.10.2	宫格导航设计	50	3.4.5 wx.navigateBack返回上一页	102
2.10.3	香哈头条初始化数据	53	3.4.6 设置导航条	103
2.10.4	香哈头条列表渲染及绑定数据	54	3.5 媒体组件	104
			3.5.1 audio音频	104
2.10.5	香哈头条模板引用	58	3.5.2 image图片	107
2.11	小结	60	3.5.3 video视频	110
			3.5.4 camera相机	112

第3章 用微信小程序组件构建UI界面 61

3.5.5 live-player实时音视频播放		113
3.5.6 live-pusher实时音视频录制		114
3.6 地图组件		115
3.7 画布组件		119
3.8 沙场大练兵：表单登录注册微信小程序		121
3.1 视图容器组件		62
3.1.1 view视图容器		62
3.1.2 scroll-view可滚动视图区域		63
3.1.3 swiper滑块视图容器		66
3.1.4 movable-view可移动视图容器		69
3.8.1 登录设计		122
3.8.2 手机号注册设计		127
3.8.3 企业用户注册设计		131
3.1.5 cover-view覆盖原生组件的视图容器		70
3.9 小结		138

第4章 必备的微信小程序API 139

3.2 基础内容组件		72
3.2.1 icon图标		72
3.2.2 text文本		73
3.2.3 progress进度条		74
3.2.4 rich-text富文本		74
4.1 请求服务器数据API		140
4.2 文件上传与下载API		142
4.2.1 wx.uploadFile文件上传		143
4.2.2 wx.downloadFile文件下载		145
3.3 丰富的表单组件		75
3.3.1 button按钮		75
3.3.2 checkbox多项选择器		77
3.3.3 radio单项选择器		78
3.3.4 input单行输入框		79
3.3.5 textarea多行输入框		82
3.3.6 label改进表单可用性		83
3.3.7 picker滚动选择器		85
3.3.8 slider滑动选择器		91
3.3.9 switch开关选择器		93
3.3.10 form表单		95
4.3 WebSocket会话API		146
4.4 图片处理API		151
4.4.1 wx.chooseImage(OBJECT)选择图片		152
4.4.2 wx.previewImage(OBJECT)预览图片		152
4.4.3 wx.getImageInfo(OBJECT)获得图片信息		153
4.4.4 wx.saveImageToPhotosAlbum保存图片到相册		154
4.5 文件操作API		155
3.4 导航组件		96
3.4.1 navigator页面链接组件		97
3.4.2 wx.navigateTo保留当前页跳转		98
3.4.3 wx.redirectTo关闭当前页跳转		99
3.4.4 跳转到tabBar页面		101
4.5.1 wx.saveFile保存文件到本地		155
4.5.2 wx.getSavedFileList获取本地文件列表		156
4.5.3 wx.getSavedFileInfo获取本地文件信息		157
4.5.4 wx.removeSavedFile删除本地文件		158

	4.5.5	wx.openDocument打开文档	159
	4.5.6	wx.getFileInfo获取文件信息	159
4.6	数据缓存API		160
	4.6.1	数据缓存到本地	160
	4.6.2	获取本地缓存数据	162
	4.6.3	移除和清理本地缓存数据	165
4.7	位置信息API		166
	4.7.1	获得位置、选择位置、打开位置	166
	4.7.2	地图组件控制	169
4.8	设备应用API		171
	4.8.1	获得系统信息	171
	4.8.2	获取网络状态	172
	4.8.3	加速度计	172
	4.8.4	罗盘	173
	4.8.5	拨打电话	174
	4.8.6	扫码	174
	4.8.7	剪贴板	175
	4.8.8	蓝牙	175
	4.8.9	屏幕亮度	179
	4.8.10	用户截屏事件	179
	4.8.11	振动	179
	4.8.12	手机联系人	180
4.9	交互反馈API		181
	4.9.1	消息提示框	181
	4.9.2	模态弹窗	183
	4.9.3	操作菜单	184
4.10	登录API		185
4.11	微信支付API		191
	4.11.1	微信小程序支付介绍	191
	4.11.2	微信小程序支付实战	193
4.12	分享API		212
4.13	沙场大练兵：仿豆瓣电影微信小程序		213
	4.13.1	电影顶部页签切换效果	214
	4.13.2	电影海报轮播效果	218
	4.13.3	电影列表方式布局	220
	4.13.4	电影详情页布局	224
	4.13.5	项目上传与预览	231
4.14	小结		232

第5章 微信小程序设计及问答　233

5.1	微信小程序设计		234
	5.1.1	突出重点，减少干扰项	234
	5.1.2	主次动作区分明显	234
	5.1.3	流程明确，避免打断	235
	5.1.4	局部加载反馈	235
	5.1.5	模态窗口加载反馈	235
	5.1.6	弹出式操作结果	236
	5.1.7	模态对话框操作结果	236
	5.1.8	结果页	237
	5.1.9	表单填写友好提示	237
5.2	微信小程序问答		238
5.3	小结		240

第二篇　综合案例应用

第6章 综合案例：仿智行火车票12306微信小程序　241

6.1	需求描述		243
6.2	设计思路及相关知识点		244
	6.2.1	设计思路	244
	6.2.2	相关知识点	244
6.3	准备工作		245
6.4	设计流程		245
	6.4.1	底部标签导航设计	245
	6.4.2	海报轮播效果设计	248
	6.4.3	火车票查询界面设计	250
	6.4.4	火车票列表设计	261
	6.4.5	个人中心界面设计	273
	6.4.6	抢票界面设计	281
	6.4.7	项目上传和预览	290
6.5	小结		291

第7章 综合案例：仿糗事百科微信小程序　292

7.1	需求描述		293
7.2	设计思路及相关知识点		294
	7.2.1	设计思路	294
	7.2.2	相关知识点	294

7.3	准备工作	294	8.2.1 设计思路	315
7.4	设计流程	295	8.2.2 相关知识点	315
	7.4.1 顶部页签菜单滑动设计	295	8.3 准备工作	316
	7.4.2 顶部页签菜单切换效果设计	297	8.4 设计流程	317
	7.4.3 糗事列表页设计	299	8.4.1 底部标签导航设计	317
	7.4.4 视频列表页设计	307	8.4.2 海报轮播效果设计	319
	7.4.5 分享设计	309	8.4.3 宫格导航设计	321
	7.4.6 项目预览	310	8.4.4 全部分类导航设计	326
7.5	小结	311	8.4.5 现金券下拉菜单筛选条件设计	332
第8章 综合案例：仿中国婚博会微信小程序		**312**	8.4.6 现金券列表页设计	335
			8.4.7 婚博会索票界面设计	341
8.1	需求描述	314	8.4.8 获知渠道弹出层设计	346
8.2	设计思路及相关知识点	315	8.5 小结	352

第一篇　微信小程序快速入门

第1章
认识微信小程序

- 微信小程序介绍
- 微信小程序开发准备
- 微信小程序开发工具的使用
- 沙场大练兵：Hello World的创建
- 小结

2016年1月9日，腾讯公司启动了微信小程序产品的研发，于2017年1月9日正式发布。微信小程序也被称为微信应用号。不同于微信订阅号或公众号，微信小程序被赋予了应用程序的能力，它是一种无须安装即可使用的应用，实现了应用"触手可及"的梦想，用户扫一扫或者搜一下即可打开应用。也体现了"用完即走"的理念，用户不再需要关心是否安装太多应用的问题。应用将无处不在，随时随地可用，无须卸载。我们一起来认识一下微信小程序吧！

1.1 微信小程序介绍

1.1.1 初识微信小程序

微信小程序（简称小程序）是一个基于去中心化而存在的平台，它没有聚合的入口，有多种进入方式。

（1）在微信中的"发现"界面，可以找到小程序的入口，如图1.1所示。

微信小程序介绍

图1.1 微信小程序入口

（2）在微信主界面下拉，会看到用过的微信小程序。

（3）给好友或者在群里分享小程序。

小程序的界面和使用方法和App类似，图1.2所示是几个已发布的常用小程序界面。

图1.2 常用微信小程序界面

用户需要下载、安装才可以使用App，安装时还会考虑App占用多大存储空间，哪些程序应该卸载掉以释放空间。微信小程序则无需安装，直接使用，不占用存储空间。用户在使用微信小程序后，可以用完即走。例如，我们去餐馆点菜，并不需要去下载这个餐馆的应用程序，只需要在餐馆扫描一下二维码，即可在小程序里点菜，之后并不需要去卸载应用程序，直接关闭小程序即可。

微信小程序看起来是程序，但它以完全不同于App的状态出现，具有更灵活的应用组织形态。

1.1.2 微信小程序的功能

小程序提供的功能如下。

（1）**分享页功能**。用户可以将小程序的当前页面分享给好友，如分享北京到上海的火车票列表页面，用户打开时是这个页面的实时数据，而不需要再次启动微信小程序。

（2）**分享对话功能**。用户可以将对话分享给好友或者微信群。

（3）**线下扫码进入微信小程序功能**。该功能提示用户附近有哪些微信小程序可以使用，扫描二维码就可以使用微信小程序。

（4）**挂起状态功能**。例如，用户使用微信小程序时可以先接电话，接完电话后可以继续使用微信小程序进行相关操作。

（5）**消息通知功能**。商户可以发送消息给接受过服务的用户，用户同时可以使用微信小程序的客服功能联系商户。

（6）**实时音视频录制播放功能**。通过此功能可以随时随地进行直播或者录播。

（7）**硬件连接功能**。通过使用NFC功能，可以把手机当成公交卡、门禁卡等使用，通过Wi-Fi连接功能，可以进行Wi-Fi连接。

（8）**小游戏功能**。利用微信小程序制作的"跳一跳"小游戏，让游戏大门从此打开，让用户知道小程序也可以制作游戏。

（9）**公众号关联功能**。微信小程序可与公众号进行关联，公众号可关联不同主体的3个小程序，还可关联同一主体的10个小程序。

（10）**搜索查找功能**。用户可以根据关键字或品牌名称查找小程序。

（11）**识别二维码功能**。用户长按识别小程序码可以进入小程序。

小程序不提供的功能如下。

（1）小程序没有集中入口，没有应用商店。

（2）小程序没有订阅关系，没有粉丝，只有访问量。

（3）小程序不能推送消息。

1.1.3 微信小程序的使用场景

从上线开始，各种小程序就如雨后春笋般出现，小程序有哪些适合的使用场景呢？在发布小程序的时候，要选择服务类目。通过这些服务类目，我们能知道小程序的使用场景。服务类目分为个人服务类目和企业服务类目。个人服务类目针对以个人为开发主体的小程序，服务范围小；企业服务类目针对以企业为开发主体的小程序，服务范围大，如表1.1、表1.2所示。

表1.1　个人服务类目

服务类目	服务范围	小程序案例
出行与交通	代驾	上海代驾服务
生活服务	家政、丽人、摄影/扩印、婚庆服务、环保回收/废品回收	掌上生活服务中心
餐饮	点评与推荐、菜谱、餐厅排队	上海餐饮

续表

服务类目	服务范围	小程序案例
旅游	出境Wi-Fi、旅游攻略	四川旅游攻略网
商业服务	会展服务、律师	律师+
快递业与邮政	邮政、装卸搬运、快递、物料	上门取快递
教育	婴幼儿教育、在线教育、特殊人群教育、教育装备、教育信息服务	全脑教育2048数字小游戏
工具	字典、图片/音频/视频、计算类、报价/比价、信息查询、效率、健康管理、企业管理、记账、日历、天气、办公、预约/报名	图文速成工具
体育	体育培训、在线健身	体育福利彩票

表1.2 企业服务类目

服务类目	服务范围	小程序案例
快递业与邮政	快递、物流、邮政、装卸搬运、仓储	申通快递小哥
教育	培训机构、教育信息服务、驾校培训、特殊人群教育、学历教育、婴幼儿教育、在线教育、教育装备、出国移民、出国留学	教育培训学院
医疗	公立医疗机构、私立医疗机构、就医服务、健康咨询/问诊、医疗保健信息服务、药品（非处方药）销售、医疗器械信息展示、药品信息展示、医疗器械生产企业、医疗器械经营/销售、互联网医院、血液中心/血库及干细胞库	海外医疗平台
政务民生	交警、交通违法、车管所服务、水电局、城市道路、高速服务、户政、治安、出入境、边防、消防、国安、司法、公证、检察院、法院、纪检审计、财政、民政、住建、公积金、党团组织、教育、社会保障、人力资源、环保、气象、工商、安监、医疗、计划生育、文化、体育、水利、食药监、质监、新闻出版及广电、税务、知识产权、旅游、信访、物价粮食、城管、监狱戒毒、海关、邮政、检验检疫、交通、商务、航空、街道居委、农林畜牧海洋、社科档案、政府应急办、科学技术与地质、统计、经济发展与改革、烟草管理单位、政务服务大厅	智行无忧
金融业	银行、信托、基金、证券/期货、证券/期货投资咨询、网络借贷信息中介（P2P）、非金融机构自营小额贷款、保险、征信业务、新三板信息服务平台、股票信息服务平台、股票信息服务平台、外汇兑换、消费金融	招行消费金融
出行与交通	打车（网约车）、顺风车（拼车）、航空、路况、地铁、水运、路桥收费、加油/充电桩、城市交通卡、城市共享交通、高速服务、火车、公交、长途客运、停车、代驾、租车	滴滴代驾
房地产	物业管理、房地产开发经营、房地产、装修/建材	玉龙房地产
生活服务	票务、生活缴费、家政、丽人、宠物（非医院类）、宠物医院/兽医、环保回收/废品回收、贵金属回收、综合生活服务平台、摄影/扩印、婚庆服务、养车/修车、休闲娱乐、搬家公司、百货/超市/便利店、洗浴保健	58同城
IT科技	硬件与设备、电信运营商、软件服务提供商	IT之家
餐饮	点评与推荐、菜谱、餐饮服务场所、餐厅排队、点餐平台、外卖平台	九鼎轩餐饮
旅游	旅游线路、旅游攻略、酒店服务、公寓/民宿、门票、签证、出境Wi-Fi、景区服务	携程旅游攻略

续表

服务类目	服务范围	小程序案例
时政信息	政治（含法律法规）、经济、军事、外交、农业、历史、社会突发事件、报社、头条、新闻、媒体、出版社	时政知识点测试
文娱	资讯、小说、视频、语音、FM/电台、音乐、图片、有声读物、动漫	文娱小咖
工具	记账、投票、日历、天气、备忘录、办公、字典、图片/音频/视频、计算类、报价/比价、信息查询、网络代理、效率、预约、健康管理、发票查询、企业管理	发票助手工具
电商平台	电商平台	京东购物
商家自营	百货、食品、酒/盐、保健品、汽车/其他交通工具的配件、图书报刊/音像/影视/游戏/动漫、成品油、纪念币发售、服装/鞋/箱包、海淘、玩具/母婴用品（不含食品）、家电/数码/手机、美妆/洗护、珠宝/饰品/眼镜/钟表、运动/户外/乐器、鲜花/园艺/工艺品、家居/家饰/家纺、汽车内饰/外饰、办公/文具、宠物/农资、五金/建材/化工/矿产品、机械/电子器件等	58商家通
商业服务	广告/设计、公关/推广/市场调查、法律服务、律所、公证、招聘/求职（中介类）、会展服务、拍卖公司（非文物）、文物拍卖公司、专利代理、亲子鉴定、典当、会计师事务所、税务师事务所、一般财务服务、农林牧渔	律师咨询中心
公益	公益慈善、基金会	腾讯公益
社交	陌生人交友、熟人交友、社区/论坛、笔记、婚恋、问答	爱社交心理
体育	体育场馆服务、体育赛事、体育培训、在线健身	雷速体育
汽车	养车/修车、汽车资讯、汽车报价/比价、车展服务、汽车经销商/4S店、汽车厂商	汽车之家

1.1.4 微信小程序能取代App吗

原生App一般要同时开发iOS和Android两版，而小程序只需要做一版。毫无疑问，这点是小程序最大的优势。从这个角度来看，小程序是"跨平台"的。

在现阶段，开发一套逻辑完整的应用程序，小程序的开发效率是低于App的。小程序独立出了一个封闭的生态。

小程序虽是跨平台的，但是缺乏成熟的组件，缺少统计、绘图组件，以前的echarts和hightcharts都无法使用。

小程序不支持WebView，大量已被静态化好的HTML页面完全没办法在小程序上展示。

小程序想取代Android和iOS还要走很长的路，是蓝海还是死海需要时间来验证。

小程序经过腾讯公司的扶持和发展，已经吸引了很多企业使用，作为与iOS、Android、公众号、网站并行的流量入口。

1.1.5 微信小程序的发展历程

微信小程序从开始研发、正式发布到推广使用，经历了以下发展时期。

（1）2016年1月9日，微信团队首次提出应用号的概念。

（2）2016年9月22日，微信公众平台对外发送小程序内测邀请，内测名额200个。

（3）2016年11月3日，微信小程序对外公测，开发完成后可以提交审核，但公测期间不能发布。

（4）2016年12月28日，张小龙在微信公开课中解答外界对微信小程序的几大疑惑，包括没有应用商店、没有推送消息等。

（5）2016年12月30日，微信公众平台对外发布公告，上线的微信小程序最多可生成10 000个带参数的二维码。

（6）2017年1月9日，微信小程序正式上线。

（7）2017年3月27日，个人开发者可以申请小程序开发和发布。

（8）2017年4月17日，小程序代码包大小限制扩大到2MB。

（9）2017年4月20日，腾讯公司发布公众号关注小程序新规则。

（10）2017年5月12日，腾讯公司发布"小程序数据助手"。

（11）2017年12月28日，微信更新的6.6.1版本开放了小游戏。

（12）2018年1月18日，微信提供了电子化的侵权投诉渠道，用户或者企业可以在微信公众平台以及微信客户端入口进行投诉。

（13）2018年1月25日，微信团队在"微信公众平台"发布公告称"从移动应用分享至微信的小程序页面，用户访问时支持打开来源应用"。

（14）2018年3月，微信正式宣布小程序广告组件启动内测，内容还包括第三方可以快速创建并认证小程序、新增小程序插件管理接口和更新基础能力，开发者可以通过小程序来赚取广告收入。

1.1.6 微信小程序带来的机会

微信小程序给很多想做程序员的人提供了机会，因为它的开发门槛很低，不需要太难的技术。学习微信小程序开发，就可以成为一名"小程序员"。例如，设计师、学生、创业者、待业青年、"网虫"、策划人员、编辑、草根站长等都可以转做程序员。

微信小程序给企业提供了流量入口，企业可以通过小程序推广自己的产品。经过腾讯公司的大力扶持，小程序已经成为各个企业非常看重的流量入口。

1.2 微信小程序开发准备

1.2.1 基础技术准备

微信小程序自定义了一套语言，称为WXML（微信标记语言），它的使用方法类似于HTML。另外，微信小程序还定义了自己的样式语言WXSS，兼容了CSS，并做了扩展；使用JavaScript来进行业务处理，兼容了大部分JavaScript功能，但仍有一些功能无法使用，所以有一定HTML、CSS、JavaScript技术功底的人学习微信小程序开发会容易很多。

微信小程序开发准备

1.2.2 开发准备

Step1：在"微信公众平台"注册微信开发者账号。单击"立即注册"，在"注册"界面选择"小程序"，在"小程序注册"界面根据提示填写相关信息完成注册。

在微信公众平台中，选择"小程序"→"小程序开发文档"，如图1.3（a）所示，可以打开帮助文档界面，如图1.3（b）所示。

在帮助文档里有介绍、设计、小程序开发、运营、数据、社区6个菜单，针对不同角色的用户提供了不同内容的帮助文档。开发人员经常会用到这里的简易教程、框架的使用、组件的介绍、API、工具以及腾讯云支持等内容。

Step2：在文档工具里，根据自己的操作系统，下载微信小程序的开发工具，如图1.4所示。

图1.3（a） 开发文档

图1.3（b） 帮助文档

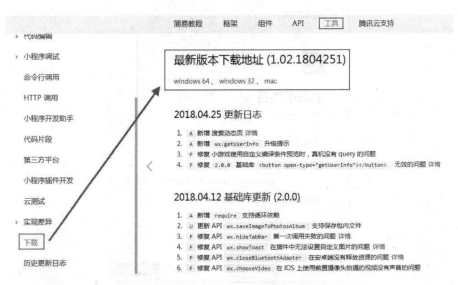

图1.4 下载开发工具

Step3：按照提示完成开发工具的安装，安装完成后用微信扫描二维码登录。开发工具提供了小程序项目和公众号网页项目两个调试类型，如图1.5所示。

图1.5　开发工具

1.3　微信小程序开发工具的使用

1.3.1　创建项目

微信小程序开发工具的使用

在开发工具里单击"小程序项目"，进入到"小程序项目"界面，可以添加一个新的项目。在这个界面里需要填写项目目录、AppID和项目名称，如图1.6所示。

图1.6　添加项目

获取微信小程序AppID，需要在"微信公众平台"中登录1.2.2节中注册的账户，在 "设置"→"开发设置"中，查看微信小程序的 AppID，如图1.7所示。

不可直接使用服务号或订阅号的 AppID。也可以不填写AppID，但功能会受限。

图1.7　获取AppID

如果要以非管理员微信号在手机上体验该小程序，还需要"绑定开发者"，即在"用户身份"→"开发者"模块，绑定需要体验该小程序的微信号。

输入AppID后，在桌面上建立一个"demo"文件夹，并将其选择为项目目录，在项目名称中输入"demo"，勾选"建立普通快速启动模板"选项（还可以选择"建立插件快速启动模板"创建插件项目），单击"确定"按钮即可，如图1.8所示。

图1.8　创建demo项目

1.3.2 开发者工具界面

创建项目后，进入到微信开发者工具界面，界面大致可以分为6个区域：①菜单栏区域，②模拟器、编辑器、调试器显示与隐藏区域，③模拟器区域，④编辑器区域，⑤调试器区域，⑥工具栏区域，如图1.9所示。

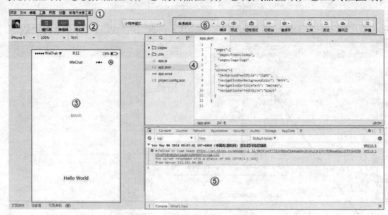

图1.9　开发者工具界面

① 菜单栏区域：包含项目、文件、编辑、工具、界面、设置、微信开发者工具菜单。

② 模拟器、编辑器、调试器显示与隐藏区域：是用来控制模拟器区域、编辑器区域、调试器区域的显示与隐藏的便捷操作按钮。

③ 模拟器区域：用来显示小程序项目的界面。

④ 编辑器区域：用来进行代码编写的区域。

⑤ 调试器区域：用来进行调试的区域。

⑥ 工具栏区域：包含编译、预览、远程调试、切后台、清缓存、上传、测试、腾讯云、详情工具栏按钮。

1.3.3 模拟器区域

模拟器区域用来显示小程序界面。在小程序开发过程中，小程序界面随着代码编写可以实时变化，方便小程序的开发和调试。同时模拟器可以模拟小程序在各个终端设备上的操作效果；可以设置小程序运行的终端设备，如iPhone 5、iPhone 6等；设置模拟器区域的百分比大小；模拟设置Wi-Fi、2G、3G等网络连接情况，如图1.10所示。

图1.10　模拟器区域

1.3.4 编辑器区域

编辑器区域分为两部分：一部分用来展示项目文件目录和文件结构；另一部分用来进行代码编辑，如图1.11所示。

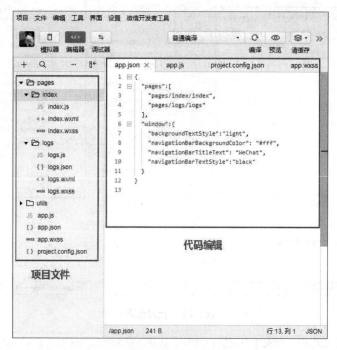

图1.11 编辑器区域

（1）在项目目录上单击鼠标右键可以在硬盘打开文件目录，对文件目录重新命名，删除目录，在该目录下查找指定内容、新建文件等，如图1.12所示。

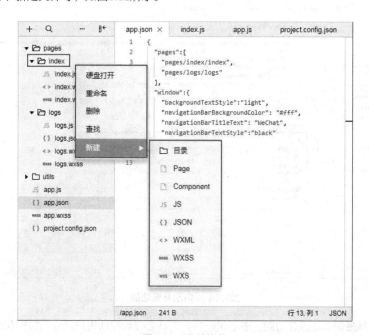

图1.12 文件操作

（2）在代码编辑区域里编写代码，可以通过模拟器区域，实时预览编辑的情况。修改 wxss、wxml 文件，会刷新当前页面（page），修改js文件或者json文件，会重新编译小程序，如图1.13所示。

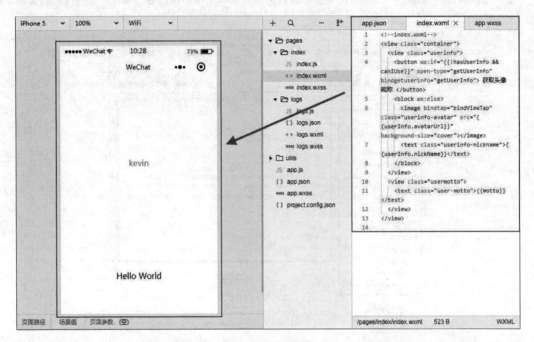

图1.13　代码编写

（3）在代码编写过程中，开发工具提供自动补全功能。在编辑js文件时，开发工具会帮助开发者补全所有的 API，并给出相关的注释解释；编辑wxml文件时，会帮助开发者直接写出相关的标签；编辑json文件时，会帮助开发者补全相关的配置，并给出实时的提示，如图1.14所示。

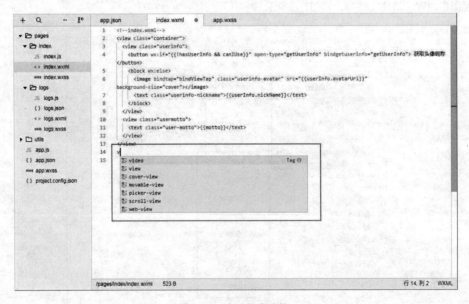

图1.14　自动补全功能

（4）开发工具提供自动保存功能，书写代码后，工具会自动帮助用户保存当前的代码编辑状态，直接关闭工具或者切换到别的项目,并不会丢失已经编辑的文件内容,但需要注意的是,只有保存文件,修改内容才会真实地写到硬盘上,并触发实时预览。

1.3.5 调试器区域

小程序的常用调试工具有:Console、Sources、Network、Storage、AppData、Wxml。除了这6个调试选项外,还有一些不常用的工具:Application为记录加载的资源信息,Security为安全和认证,Audits为性能诊断和优化建议,Sensor用来选择模拟地理位置,Trace为性能监测数据,在这里不做详细介绍。

(1)Console窗口用来显示小程序的错误输出信息和调试代码,除了可以输出错误信息,还可以进行代码编写和调试,如图1.15所示。

图1.15 Console功能

(2)Sources窗口用于显示当前项目的脚本文件,在 Sources中开发者看到的文件是经过处理之后的脚本文件,开发者的代码都会被包裹在 define 函数中,并且对于 Page代码,有require 的主动调用,如图1.16所示。

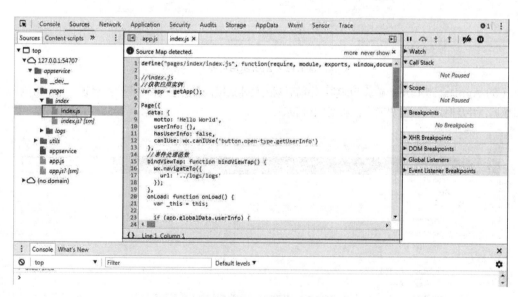

图1.16 Sources功能

(3)Network用来观察发送的请求和调用文件的信息,包括文件名称、路径、大小、调用的状态、时间等,如图1.17所示。

(4)Storage窗口用于显示当前项目,使用wx.setStorage 或者 wx.setStorageSync 后的数据存储情况,如图1.18所示。

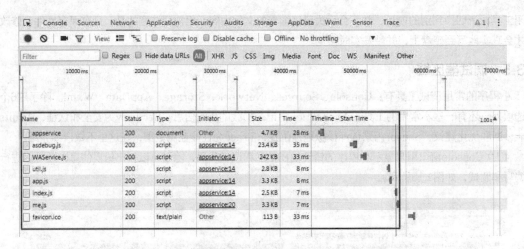

图1.17　Network功能

图1.18　Storage功能

（5）AppData窗口用于显示当前项目当前时刻的具体数据，实时地反馈项目数据情况。用户可以在此处编辑数据，并及时地反馈到界面上，如图1.19所示。

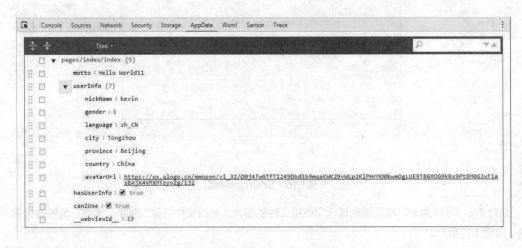

图1.19　AppData功能

（6）Wxml窗口用于帮助开发者开发Wxml转化后的界面。在这里可以看到真实的页面结构以及结构对应的 wxss 属性，同时可以修改对应的 wxss 属性，如图1.20所示。

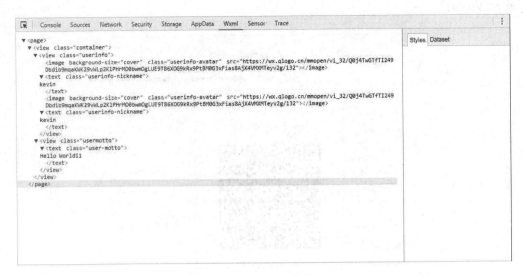

图1.20　Wxml功能

1.3.6　工具栏区域

1. 编译操作

我们可以通过编译按钮或者使用快捷键 Ctrl + B编译当前小程序的代码，并自动刷新模拟器。为了方便调试，开发者还可以添加或选择已有的自定义编译条件进行编译和代码预览，如图1.21所示。

图1.21　编译

2. 预览

单击预览按钮，可以将小程序上传，生成二维码，通过扫描二维码可以在手机上预览小程序，如图1.22所示。

3. 前后台切换

工具栏中的前后台切换按钮可以帮助开发者模拟一些客户端的操作环境。例如，在操作微信小程序过程中,突然进来电话,如果接电话,小程序就会从前台进入到后台,重新访问小程序时,又会从后台进入到前台,如图1.23所示。

图1.22 预览(本图中二维码只是示意,请扫描自己操作生成的二维码)

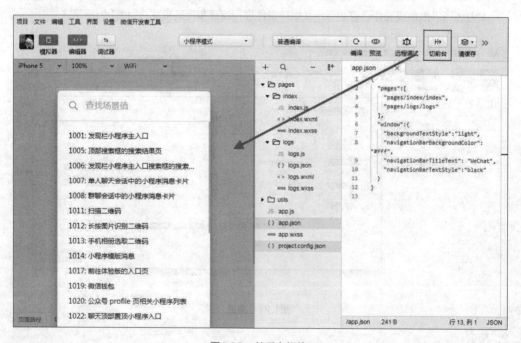

图1.23 前后台切换

4. 清缓存

清缓存包括清除数据缓存、清除文件缓存、清除授权数据、清除网络缓存、清除登录状态、全部清除功能,如图1.24所示。

图1.24 清缓存

5. 上传、测试

小程序开发完成后，需要上传到腾讯服务器进行测试，然后可以获取测试报告，根据测试报告进行相应的修改，如图1.25、图1.26所示。

图1.25 上传

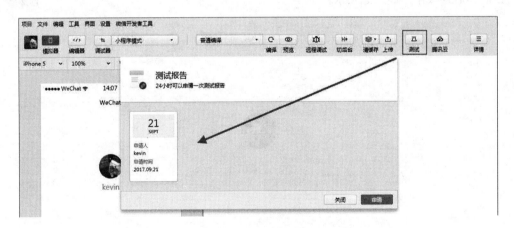

图1.26 测试报告申请

1.3.7 常用快捷键

1. 格式调整快捷键

Ctrl+S：保存文件。

Ctrl+[，Ctrl+]：代码行缩进。

Ctrl+Shift+[，Ctrl+Shift+]：折叠打开代码块。

Ctrl+C，Ctrl+V：复制粘贴，如果没有选中任何文字则复制粘贴一行。

Shift+Alt+F：代码格式化。

Alt+Up，Alt+Down：上下移动一行。

Shift+Alt+Up，Shift+Alt+Down：向上向下复制一行。

Ctrl+Shift+Enter：在当前行上方插入一行。

Ctrl+Shift+F：全局搜索。

Ctrl+B：可以编译当前代码，并自动刷新模拟器。

2. 光标相关快捷键

Ctrl+End：移动到文件结尾。

Ctrl+Home：移动到文件开头。

Ctrl+I：选中当前行。

Shift+End：选择从光标到行尾。

Shift+Home：选择从行首到光标处。

Ctrl+Shift+L：选中所有匹配。

Ctrl+D：选中匹配。

Ctrl+U：光标回退。

3. 界面相关快捷键

Ctrl + \：隐藏侧边栏。

Ctrl + M：打开或者隐藏模拟器。

1.4 沙场大练兵：Hello World的创建

在创建项目之后，开发工具会添加默认的目录和页面，在默认的页面上，可以看到有"Hello World"文字，如图1.27所示。

Hello World的创建

图1.27 Hello World界面

下面,我们分析一下Hello World是怎么创建出来的。

(1)在pages/index/index.js文件里,Page的data中提供数据源motto,data的数据可以动态地绑定到WXML页面中,如图1.28所示。

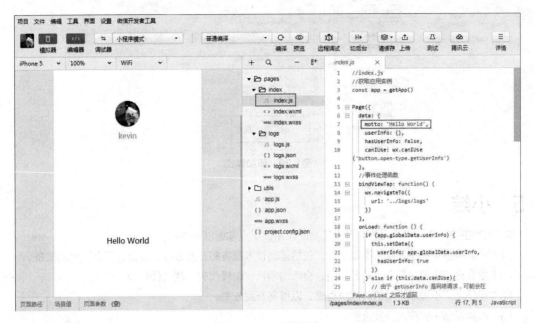

图1.28　motto数据源

(2)在pages/index/index.wxml文件里,通过双大括号({{}})的方式,将motto绑定到页面里,motto对应的值就可以在页面里显示出来,如图1.29所示。

图1.29　绑定motto

(3)在pages/index/index.wxss文件里,通过class的方式给Hello World添加样式,距顶部的高度200px,如图1.30所示。

在实际的开发过程中也是这样来进行的,在js文件里进行业务逻辑的处理,动态地提供数据;在wxml文件里绑定数据,渲染界面;在wxss文件里添加样式,美化页面,就可以完成微信小程序的开发了。

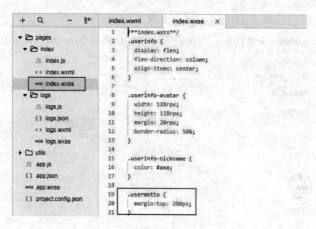

图1.30 添加样式

1.5 小结

本章内容主要认识微信小程序和开发工具的使用，重点掌握以下内容。

（1）做好微信小程序开发的准备工作，包括基础技术准备和开发账号、文档、开发工具的准备。

（2）学会微信小程序开发工具的使用，会添加项目、编辑代码、调试代码等。

（3）记住微信小程序常用的一些快捷键，以提高开发效率。

（4）理解微信小程序的开发流程。

第2章
微信小程序框架分析

- 微信小程序目录结构介绍
- 微信小程序注册程序应用
- 微信小程序注册页面的使用
- 微信小程序如何绑定数据
- 微信小程序条件渲染
- 微信小程序列表渲染
- 微信小程序定义模板
- 微信小程序的引用功能
- WXS小程序脚本语言
- 沙场大练兵：仿香哈菜谱微信小程序
- 小结

微信小程序框架是进行微信小程序开发必先理解的内容。微信小程序框架，让开发者在微信中通过简单、高效的方式开发具有原生App体验的服务。微信小程序框架分为逻辑层和视图层，逻辑层用来处理业务逻辑，而视图层用来渲染页面。视图层描述语言WXML和视图样式WXSS，再加上JavaScript逻辑层语言和json配置文件，构筑起了微信小程序框架。本章我们一起来分析微信小程序的框架。

2.1 微信小程序目录结构介绍

微信小程序目录结构可以分为3个部分——框架全局文件、工具类文件和框架页面文件,如图2.1所示。

微信小程序目录结构介绍

图2.1 微信小程序框架目录

2.1.1 框架全局文件

框架全局文件必须放在项目的根目录中。框架全局文件包括4个文件:app.js小程序逻辑文件(定义全局数据以及定义函数文件)、app.json小程序公共设置文件、app.wxss小程序公共样式表、project.config.json小程序项目个性化配置文件。它们对所有页面都有效,如表2.1所示。

表2.1 框架全局文件

文件	是否必填	作用
app.js	是	装载小程序逻辑
app.json	是	装载小程序公共设置
app.wxss	否	装载小程序公共样式
project.config.json	是	装载小程序项目个性化配置

1. app.js小程序逻辑文件

app.js文件用来定义全局数据和函数的使用,它可以指定微信小程序的生命周期函数。生命周期函数可以理解为微信小程序自己定义的函数,例如onLaunch(监听小程序初始化)、onShow(监听小程序显示)、onHide(监听小程序隐藏)等,在不同阶段不同场景可以使用不同的生命周期函数。此外,app.js中还可以定义一些全局的函数和数据,其他页面引用app.js文件后就可以直接使用,如图2.2所示。

2. app.json小程序公共设置文件

app.json文件可以对5个功能进行设置:配置页面路径、配置窗口表现、配置标签导航、配置网络超时、配置debug模式。具体如图2.3所示。

(1)配置页面路径。页面路径定义了一个数组,存放多个页面的访问路径,它是进行页面访问的必要条件。如果在这里没有配置页面访问路径,页面被访问时就会报错;在这里定义了页面访问路径,微信小程序框架就可以在页面文件夹下建立相应名称的文件夹以及文件,免去用户手动添加文件夹和文件的痛苦,如图2.4所示。

```
App({
  onLaunch: function () {          // ← 生命周期函数
    // 展示本地存储能力
    var logs = wx.getStorageSync('logs') || []
    logs.unshift(Date.now())
    wx.setStorageSync('logs', logs)
    // 登录
    wx.login({                     // ← 登录操作
      success: res => {
        // 发送 res.code 到后台换取 openId, sessionKey, unionId
      }
    })
    // 获取用户信息
    wx.getSetting({                // ← 获取用户信息
      success: res => {
        if (res.authSetting['scope.userInfo']) {
          // 已经授权，可以直接调用 getUserInfo 获取头像昵称，不会弹框
          wx.getUserInfo({
            success: res => {
              // 可以将 res 发送给后台解码出 unionId
              this.globalData.userInfo = res.userInfo
              // 由于 getUserInfo 是网络请求，可能会在 Page.onLoad 之后才返回
              // 所以此处加入 callback 以防止这种情况
              if (this.userInfoReadyCallback) {
                this.userInfoReadyCallback(res)
              }
            }
          })
        }
      }
    })
  },
  globalData: {                    // ← 定义全局数据
    userInfo: null
  }
})
```

图2.2 app.js小程序逻辑

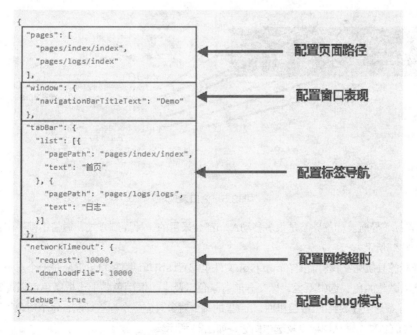

图2.3 app.json的5个功能

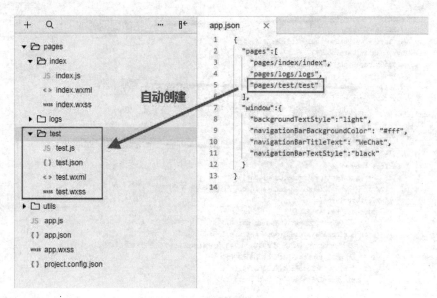

图2.4 自动创建页面

（2）配置窗口表现。窗口表现用于配置小程序的状态栏、导航条、标题和窗口背景色，可以设置导航条背景色（navigationBarBackgroundColor）、导航条文字（navigationBarTitleText）以及导航条文字颜色（navigationBarTextStyle）；还可以设置窗口是否可以下拉刷新（enablePullDownRefresh）（默认值是不可以下拉刷新的），设置窗口的背景色（backgroundColor）和下拉背景字体或者loading样式（backgroundTextStyle），如图2.5所示。

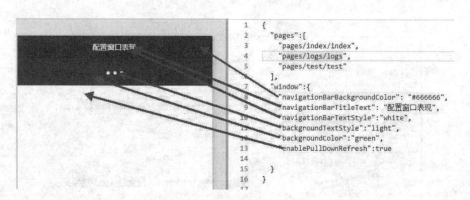

图2.5 窗口表现

（3）配置标签导航。标签导航是很多移动App都会采用的一种导航方式，微信小程序同样可以实现这样的效果，如图2.6所示。

怎么制作标签导航呢？我们需要在app.json文件里配置tabBar属性。tabBar是一个对象，它可以配置标签导航文字的默认颜色、选中颜色，标签导航背景色以及上边框颜色。上边框颜色可以配置两种颜色：black/white。标签导航存放到list数组里面，list里的每个对象对应一个标签导航，每个对象里可以配置标签导航的路径、导航名称、默认图标以及选中图标，如图2.7所示。

（4）配置网络超时。可以配置网络请求、文件上传、文件下载时最大的请求时间，超过这个时间，则不再请求。

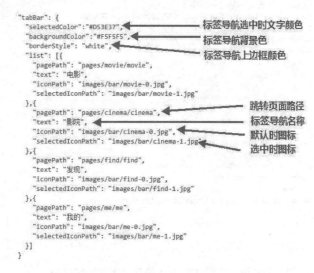

图2.6　猫眼电影App标签导航　　　　　图2.7　猫眼电影微信小程序标签导航配置

（5）配置debug模式。配置debug模式可方便微信小程序开发者调试开发程序，如图2.8和图2.9所示为没有开启debug模式和开启debug模式的调试信息对比。

图2.8　没有开启debug模式

从图2.8和图2.9可以看出，开启debug模式后，可以看到每一步的调用情况、访问路径以及错误信息，这样更加方便开发者进行调试工作。

app.json作为全局配置文件就是提供配置页面路径、配置窗口的表现、配置底部标签导航、配置网络连接超时、配置debug模式这些功能，配置也比较容易。

3. app.wxss小程序公共样式表

app.wxss文件对CSS样式进行了扩展，和CSS的使用方式一样，类选择器和行内样式的写法兼容大部分CSS样式，有一些CSS样式在这里是不起作用的。app.wxss还形成了自己的风格，是对所有页面定义的一个全局样式。只要页面有全局样式里的class，就可以渲染全局样式里的效果；但如果页面又重新定义了这个

class样式，则会把全局的覆盖掉，使用自己的样式，如图2.10所示。

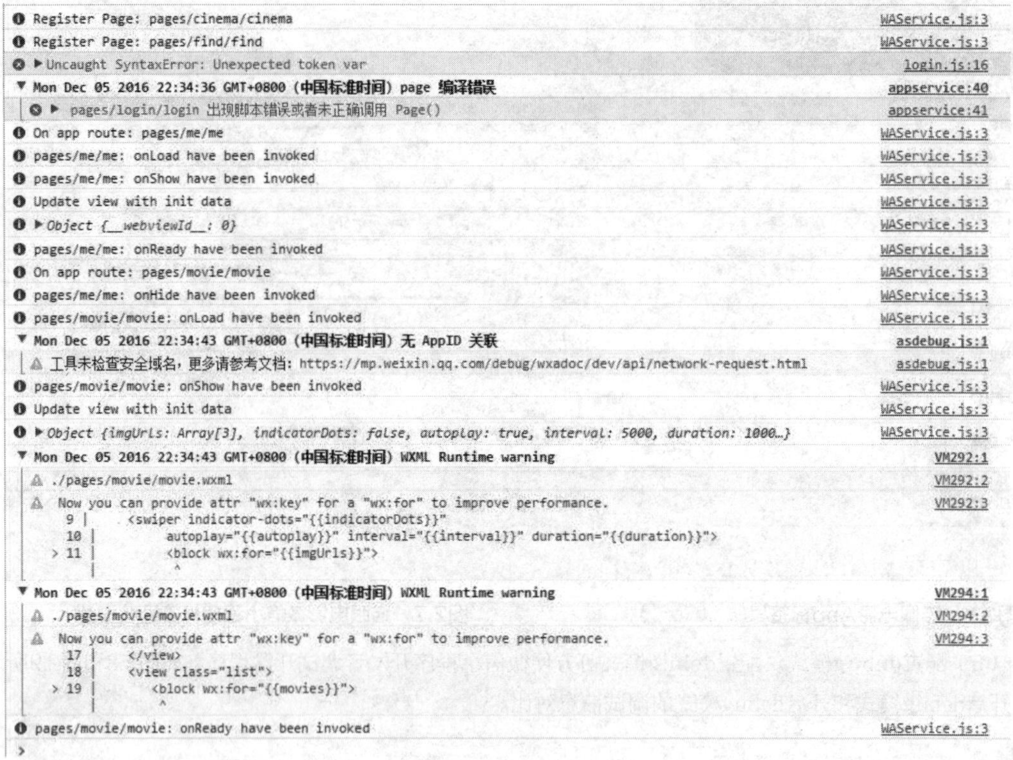

```
ⓘ Register Page: pages/cinema/cinema                                              WAService.js:3
ⓘ Register Page: pages/find/find                                                  WAService.js:3
ⓧ ▶ Uncaught SyntaxError: Unexpected token var                                      login.js:16
▼ Mon Dec 05 2016 22:34:36 GMT+0800 (中国标准时间) page 编译错误                    appservice:40
  ⓧ ▶ pages/login/login 出现脚本错误或者未正确调用 Page()                          appservice:41
ⓘ On app route: pages/me/me                                                       WAService.js:3
ⓘ pages/me/me: onLoad have been invoked                                           WAService.js:3
ⓘ pages/me/me: onShow have been invoked                                           WAService.js:3
ⓘ Update view with init data                                                      WAService.js:3
ⓘ ▶ Object {__webviewId__: 0}                                                     WAService.js:3
ⓘ pages/me/me: onReady have been invoked                                          WAService.js:3
ⓘ On app route: pages/movie/movie                                                 WAService.js:3
ⓘ pages/me/me: onHide have been invoked                                           WAService.js:3
ⓘ pages/movie/movie: onLoad have been invoked                                     WAService.js:3
▼ Mon Dec 05 2016 22:34:43 GMT+0800 (中国标准时间) 无 AppID 关联                     asdebug.js:1
  ⚠ 工具未检查安全域名，更多请参考文档：https://mp.weixin.qq.com/debug/wxadoc/dev/api/network-request.html   asdebug.js:1
ⓘ pages/movie/movie: onShow have been invoked                                     WAService.js:3
ⓘ Update view with init data                                                      WAService.js:3
ⓘ ▶ Object {imgUrls: Array[3], indicatorDots: false, autoplay: true, interval: 5000, duration: 1000...}   WAService.js:3
▼ Mon Dec 05 2016 22:34:43 GMT+0800 (中国标准时间) WXML Runtime warning             VM292:1
  ⚠ ./pages/movie/movie.wxml                                                      VM292:2
  ⚠ Now you can provide attr "wx:key" for a "wx:for" to improve performance.
      9 |      <swiper indicator-dots="{{indicatorDots}}"                          VM292:3
     10 |        autoplay="{{autoplay}}" interval="{{interval}}" duration="{{duration}}">
    > 11 |        <block wx:for="{{imgUrls}}">
▼ Mon Dec 05 2016 22:34:43 GMT+0800 (中国标准时间) WXML Runtime warning             VM294:1
  ⚠ ./pages/movie/movie.wxml                                                      VM294:2
  ⚠ Now you can provide attr "wx:key" for a "wx:for" to improve performance.
     17 |      </view>                                                             VM294:3
     18 |      <view class="list">
    > 19 |        <block wx:for="{{movies}}">
ⓘ pages/movie/movie: onReady have been invoked                                    WAService.js:3
>
```

图2.9　开启debug模式

除了app.wxss提供的默认全局样式，用户自己也可以定义一些全局样式，这样方便每个页面的使用，又不用在每个页面都写一次，达到一次定义，其他页面直接引用的复用效果。

4. project.config.json小程序项目个性化配置文件

在使用微信小程序开发者工具时，开发者都会针对各自喜好做一些个性化配置，例如界面颜色、编译配置等。当换了另外一台计算机重新安装工具的时候，用户还要重新配置。因此，小程序开发者工具在每个项目的根目录都会生成一个 project.config.json文件，用户在工具上做的任何配置都会写入这个文件。重新安装工具或者换计算机工作时，用

```
/**app.wxss**/
.container {
  height: 100%;
  display: flex;
  flex-direction: column;
  align-items: center;
  justify-content: space-between;
  padding: 200rpx 0;
  box-sizing: border-box;
}
```

图2.10　小程序公共样式表

户只要载入同一个项目的代码包，开发者工具就会自动恢复到当时开发项目时的个性化配置，其中包括编辑器的颜色、代码上传时自动压缩等一系列选项。

2.1.2　工具类文件

在微信小程序框架目录里有一个utils文件夹，它用来存放工具栏的js函数，例如可以放置一些日期格式化的函数、时间格式化的函数等一些常用的函数。定义完这些函数后，要通过module.exports将定义的函数名称注册进来，在其他的页面才可以使用，图2.11所示为时间格式化工具类文件。

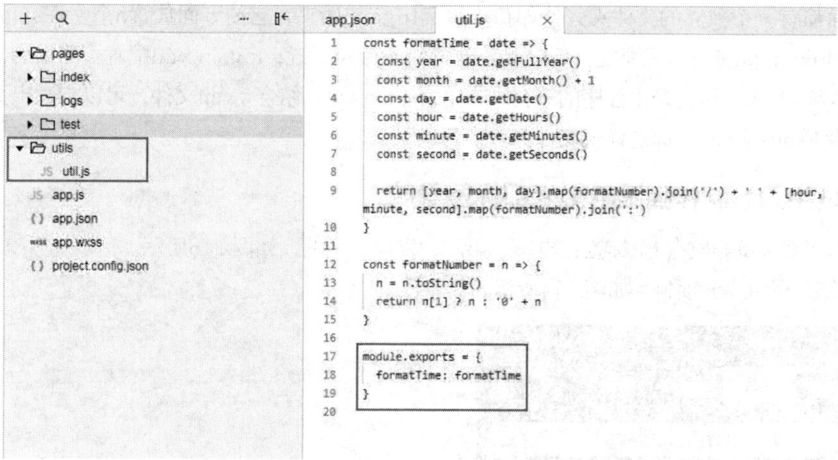

图2.11 utils.js工具类文件

2.1.3 框架页面文件

一个小程序框架页面文件由5个文件组成，分别是js页面逻辑、json页面配置、wxml页面结构、wxs小程序脚本语言、wxss页面样式表，如表2.2所示。

表2.2 框架页面文件

文件类型	是否必填	作用
js	是	页面逻辑
json	否	页面配置
wxml	是	页面结构
wxs	否	小程序脚本语言
wxss	否	页面样式表

微信小程序的框架页面文件，都放置在pages文件夹下面，如图2.12所示。

图2.12 页面文件

每个页面都有一个独立的文件夹,比如日志页面logs文件夹,它的下面放置5个文件:logs.js进行业务路径处理;logs.json进行页面配置,可以覆盖全局App.json配置;logs.wxml配置页面结构,负责渲染页面;WXS(WeiXin Script)是小程序的一套脚本语言,logs.wxs结合wxml文件,可以构建出页面的结构;logs.wxss是针对logs.wxml页面的样式文件。

2.1.4　小试牛刀:制作猫眼电影底部标签导航

猫眼电影底部标签导航有4个标签:电影、影院、发现、我的,如图2.13所示。

(1)新建一个movie项目,如图2.14所示。

图2.13　猫眼电影底部标签导航

图2.14　添加项目

(2)将准备好的底部标签导航图标拷贝到movie项目下面。

(3)打开App.json配置文件,在pages数组里添加4个页面路径——电影"pages/movie/movie"、影院"pages/cinema/cinema"、发现"pages/find/find"、我的"pages/me/me",保存后会自动生成相应的页面文件夹;删除"pages/index/index""pages/logs/logs"页面路径以及对应的文件夹,如图2.15所示。

(4)在window数组里配置窗口导航背景颜色为红色(#D53E37),导航栏文字为电影,字体颜色设置为白色(white),具体配置如图2.16所示。

(5)在tabBar对象里配置底部标签导航背景色为灰色(#f5f5f5),文字默认颜色为白色(white),文字选中时为红色(#D53E37),在list数组里配置底部标签导航对应的页面、导航名称、默认时图标、选中时图标,具体配置如图2.17所示。

这样就完成了猫眼电影底部标签导航的配置,单击不同的导航标签,可以切换显示不同的页面,同时导航图标和导航文字会呈现为选中状态,如图2.18所示。

第2章 微信小程序框架分析

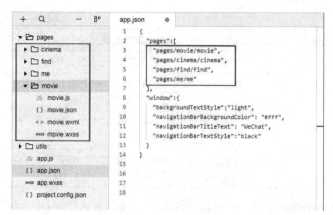

图2.15 配置页面路径

图2.16 窗口及导航栏配置

pages/movie/movie.wxml

图2.17 底部标签导航配置

图2.18 电影界面

2.2 微信小程序注册程序应用

app.js文件不仅可以定义全局函数和数据,还可以注册小程序。在App()函数里可以完成小程序的注册,并指定其生命周期函数。表2.3所示为生命周期函数的定义。

微信小程序
注册程序应用

表2.3 生命周期函数

属性	类型	描述	触发时机
onLaunch	Function	监听小程序初始化	当小程序初始化完成时,会触发onLaunch(全局只触发一次)
onShow	Function	监听小程序显示	当小程序启动,或从后台进入前台显示,会触发onShow

续表

属性	类型	描述	触发时机
onHide	Function	监听小程序隐藏	当小程序从前台进入后台，会触发onHide
onError	Function	错误监听函数	当小程序发生脚本错误，或者API调用失败时，会触发 onError 并附带错误信息
onPageNotFound	Function	页面不存在监听函数	当小程序出现要打开的页面不存在的情况，会附带页面信息回调该函数
其他	Any		开发者可以添加任意的函数或数据到Object参数中，用this可以使用

（1）onLaunch生命周期函数，用来监听小程序初始化，一旦初始化完成，就会触发该函数，这个生命周期函数只会触发一次。

（2）onShow生命周期函数，用来监听小程序显示，微信小程序有前后台定义，当用户单击左上角关闭、按Home键关闭或者突然来电话，微信小程序都没有销毁，而是进入后台，当再次进入微信或者小程序的时候就会触发onShow这个函数。只要程序启动或者从后台进入前台都会触发该函数。

（3）onHide生命周期函数，用来监听小程序隐藏，一旦微信小程序从前台进入后台，就会触发该函数。

（4）onError生命周期函数，用来监听小程序脚本或者API是否发生错误，发生错误时返回错误信息。

（5）onPageNotFound生命周期函数，当要打开的页面不存在时，会回调这个监听函数。

示例代码：

```
App({
  onLaunch: function() {
    // Do something initial when launch.
  },
  onShow: function() {
    // Do something when show.
  },
  onHide: function() {
    // Do something when hide.
  },
  onError: function(msg) {
    console.log(msg)
  },
  globalData: 'I am global data'
})
```

在页面里调用app.js全局数据：

在页面js文件里，按如下所示方法，就可以调用到全局数据globalData。

```
var AppInstance = getApp()
console.log(AppInstance.globalData)
```

不仅可以调用全局数据，还可以调用自定义的全局函数，但是不要调用生命周期函数。

（1）App() 必须在 App.js 中注册，且不能注册多个。

（2）不要在定义于 App() 内的函数中调用 getApp()，使用 this 就可以获取 App 实例。

（3）不要在 onLaunch 的时候调用 getCurrentPage()，此时 page 还没有生成。

（4）通过 getApp() 获取实例之后，不要私自调用生命周期函数。

2.3 微信小程序注册页面的使用

微信小程序注册页面的使用

在每个页面文件夹里，都有一个页面对应的js文件，比如日志logs文件夹，对应的就是logs.js文件，这个文件里的Page()函数用来注册页面。接受一个object 参数，其指定页面的初始数据、生命周期函数、事件处理函数等页面的所有业务逻辑处理都放在这个文件里。object参数如表2.4所示。

表2.4　object参数说明

属性	类型	描述
data	Object	页面的初始数据
onLoad	Function	生命周期函数—监听页面加载
onReady	Function	生命周期函数—监听页面初次渲染完成
onShow	Function	生命周期函数—监听页面显示
onHide	Function	生命周期函数—监听页面隐藏
onUnload	Function	生命周期函数—监听页面卸载
onPullDownRefresh	Function	页面相关事件处理函数—监听用户下拉动作
onReachBottom	Function	页面上拉触底事件的处理函数
onShareAppMessage	Function	用户单击右上角分享
onPageScroll	Function	页面滚动触发事件的处理函数
onTabItemTap	Function	当前是 tab 页时，点击 tab 时触发
其他	Any	开发者可以添加任意的函数或数据到 object 参数中，在页面的函数中用 this 可以访问

Page()函数使用代码如下：

```
Page({
  data: {
    text: "This is page data."
  },
  onLoad: function(options) {
    // Do some initialize when page load.
  },
  onReady: function() {
    // Do something when page ready.
  },
  onShow: function() {
    // Do something when page show.
  },
  onHide: function() {
    // Do something when page hide.
  },
  onUnload: function() {
    // Do something when page close.
  },
  onPullDownRefresh: function() {
```

```
      // Do something when pull down.
    },
    onReachBottom: function() {
      // Do something when page reach bottom.
    },
    onShareAppMessage: function () {
      // return custom share data when user share.
    },
    // Event handler.
    viewTap: function() {
      this.setData({
        text: 'Set some data for updating view.'
      })
    },
    customData: {
      hi: 'MINA'
    }
  })
```

2.3.1 页面初始化数据

data为页面初始化数据，初始化数据将作为页面的第一次渲染。data 将会以 JSON 的形式由逻辑层传至渲染层，所以其数据必须是可以转成 JSON 的格式：字符串、数字、布尔值、对象或数组。渲染界面可以通过 WXML 对数据进行绑定。

示例代码：

```
<text class="user-motto">{{motto}}</text>
Page({
  data: {
    motto: 'Hello World',
    userInfo: {}
  } })
```

2.3.2 生命周期函数

（1）onLoad页面加载：一个页面只会调用一次，接收页面参数可以获取wx.navigateTo和wx.redirectTo及<navigator/>中的 query。

（2）onShow页面显示：每次打开页面都会调用一次。

（3）onReady页面初次渲染完成：一个页面只会调用一次，代表页面已经准备妥当，可以和视图层进行交互，对界面的设置如wx.setNavigationBarTitle请在onReady之后设置。

（4）onHide页面隐藏：当调用navigateTo或底部tab切换时调用。

（5）onUnload页面卸载：当调用redirectTo或navigateBack的时候调用。

2.3.3 页面相关事件处理函数

（1）onPullDownRefresh下拉刷新：监听用户下拉刷新事件，需要在config的window选项中开启enablePullDownRefresh。当处理完数据刷新后，wx.stopPullDownRefresh可以停止当前页面的下拉刷新。

（2）onShareAppMessage用户分享：只有定义了此事件处理函数，右上角菜单才会显示"分享"按钮，用户点击"分享"按钮的时候会调用此函数，此事件需要返回一个 Object参数，用于自定义分享内容。Object参数说明如表2.5所示。

表2.5　分享参数

字段	说明	默认值
title	分享标题	当前小程序名称
desc	分享描述	当前小程序描述
path	分享路径	当前页面 path，必须是以 / 开头的完整路径

示例代码：

```
Page({
  onShareAppMessage: function () {
    return {
      title: '自定义分享标题',
      desc: '自定义分享描述',
      path: '/page/user?id=123'
    }
  }
})
```

2.3.4　页面路由管理

微信小程序的页面路由都是由微信小程序框架来管理的，框架以栈的形式维护了所有页面。栈作为一种数据结构，是一种只能在一端进行插入和删除操作的特殊线性表。它按照后进先出的原则存储数据，先进入的数据被压入栈底，最后进入的数据在栈顶，需要读数据的时候从栈顶开始弹出数据（最后一个数据被第一个读出来）。

微信小程序页面交互也是通过栈来完成的，微信小程序初始化时，新页面入栈；打开新页面时，新页面入栈；页面重定向时，当前页面出栈，新页面入栈；页面返回时，页面不断出栈，直到新页面入栈；tab切换时，页面全部出栈，只留下新的tab页面；重新加载时，页面全部出栈，只留下新的页面。

对于路由的触发方式以及页面生命周期函数如表2.6所示。

表2.6　路由的触发方式及页面生命周期函数

页面路由方式	触发时机	路由后页面	路由前页面
初始化	小程序打开的第一个页面	onLoad, onShow	
打开新页面	调用 API wx.navigateTo 或使用组件 <navigator open-type="navigate"/>	onLoad, onShow	onHide
页面重定向	调用 API wx.redirectTo 或使用组件 <navigator open-type="redirect"/>	onLoad, onShow	onUnload
页面返回	调用 API wx.navigateBack 或用户按左上角返回按钮	onShow	onUnload（多层页面返回，每个页面都会按顺序触发onUnload）
tab 切换	Function	调用 API wx.switchTab 或使用组件 <navigator open-type="switchTab"/> 或用户切换 tab	

navigateTo, redirectTo 只能打开非 tabBar 页面；switchTab 只能打开 tabBar 页面；reLaunch 可以打开任意页面；页面底部的 tabBar 由页面决定，即只要是定义为 tabBar 的页面，底部都有 tabBar；调用页面路由带的参数可以在目标页面的 onLoad 中获取。

2.3.5 自定义函数

除了初始化数据和生命周期函数，Page 中还可以定义一些特殊的函数：事件处理函数。在渲染层的组件中可以加入事件绑定，当达到触发事件时，就会执行 Page 中定义的事件处理函数。

示例代码：

```
<view bindtap="clickMe"> click me </view>

Page({
  clickMe: function() {
    console.log('view tap')
  }
})
```

2.3.6 setData设值函数

Page.prototype.setData()设值函数用于将数据从逻辑层发送到视图层，同时改变对应的 this.data 的值。

setData()参数格式：接受一个对象，以key，value的形式表示将this.data中的key对应的值改变成value。

其中 key 非常灵活，以数据路径的形式给出，如 array[2].message，a.b.c.d，并且不需要在 this.data 中预先定义。

示例代码：

```
<!--index.wxml-->
<view>{{text}}</view>
<button bindtap="changeText"> Change normal data </button>
<view>{{array[0].text}}</view>
<button bindtap="changeItemInArray"> Change Array data </button>
<view>{{object.text}}</view>
<button bindtap="changeItemInObject"> Change Object data </button>
<view>{{newField.text}}</view>
<button bindtap="addNewField"> Add new data </button>

//index.js
Page({
  data: {
    text: 'init data',
    array: [{text: 'init data'}],
    object: {
      text: 'init data'
    }
  },
  changeText: function() {
    // this.data.text = 'changed data'   // bad, it can not work
    this.setData({
      text: 'changed data'
    })
  },
  changeItemInArray: function() {
    // you can use this way to modify a danamic data path
    this.setData({
      'array[0].text':'changed data'
```

```
    })
  },
  changeItemInObject: function(){
    this.setData({
      'object.text': 'changed data'
    });
  },
  addNewField: function() {
    this.setData({
      'newField.text': 'new data'
    })
  }
})
```

 直接修改 this.data 无效，无法改变页面的状态，还会造成数据不一致。单次设置的数据不能超过1 024kB，请尽量避免一次性设置过多的数据。

2.4 微信小程序如何绑定数据

WXML页面里的动态数据，都是来自js文件Page的data，数据绑定就是通过双大括号（{{}}）将变量包起来，在WXML页面里将数据值显示出来。

示例代码如下：

微信小程序如何绑定数据

```
index.wxml
<view> {{ message }} </view>
index.js
Page({
  data: {
    message: 'Hello MINA!'
  }
})
```

2.4.1 组件属性绑定

组件属性绑定，是将data里的数据绑定到微信小程序的组件上，示例代码如下：

```
<view id="item-{{id}}"> </view>
Page({
  data: {
    id: 0
  }
})
```

2.4.2 控制属性绑定

控制属性绑定用来进行if语句条件判断，如果满足条件，则执行，否则不执行，示例代码如下：

```
<view wx:if="{{condition}}"> </view>
Page({
  data: {
    condition: true
  }
```

})
```

### 2.4.3 关键字绑定

关键字绑定常用于组件的一些关键字，像复选框组件一样，checked关键字如果等于true，则代表复选框选中；如果等于false，则代表不选中复选框，示例代码如下：

```
<checkbox checked="{{false}}"> </checkbox>
```

不要直接写 checked="false"，其计算结果是一个字符串，转成 boolean 类型后代表真值。

2.4.4 运算

可以在 {{}} 内进行简单的运算，小程序支持以下几种方式进行运算：

1. 三元运算

```
<view hidden="{{flag ? true : false}}"> Hidden </view>
```

2. 数学运算

```
<view> {{a + b}} + {{c}} + d </view>

Page({
  data: {
    a: 1,
    b: 2,
    c: 3
  }
})
```

view中的内容为 3 + 3 + d。

3. 逻辑判断

```
<view wx:if="{{length > 5}}"> </view>
```

4. 字符串运算

```
<view>{{"hello" + name}}</view>
Page({
  data:{
    name: 'MINA'
  }
})
```

5. 数据路径运算

```
<view>{{object.key}} {{array[0]}}</view>
Page({
  data: {
    object: {
      key: 'Hello'
    },
    array: ['MINA']
  }
})
```

2.4.5 小试牛刀：天气微信小程序

天气微信小程序，用来显示温度、最低温度、最高温度以及其他天气情况，下面通过数据绑定的方式，来显示天气情况，如图2.19所示。

（1）创建一个weather项目，如图2.20所示。
```

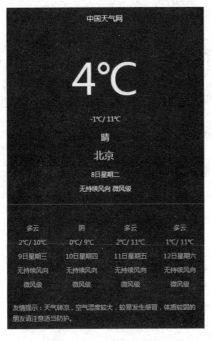
图2.19 天气微信小程序

图2.20 weather项目

（2）进入index.wxml、index.js、index.wxss文件，清空所有的内容，进入App.json，修改导航栏标题为"中国天气网"。

（3）进入index.wxml，进行当天天气情况的界面布局，包括温度、最低温度和最高温度、天气情况、城市、星期、风向情况，代码如下：

```
<view class="content">
  <view class="today">
    <view class="info">
      <view class="temp">℃</view>
      <view class="lowhigh"></view>
      <view class="type"></view>
      <view class="city"></view>
      <view class="week"></view>
      <view class="weather"></view>
    </view>
  </view>
</view>
```

（4）进入index.js，在data里提供天气的数据，让这些数据在界面里显示出来，代码如下：

```
Page({
    data:{
        temp:"4",
        low:"-1℃",
        high:"10℃",
        type:"晴",
        city:"北京",
        week:"星期二",
        weather:"无持续风向微风级"
    }
})
```

（5）进入index.wxml，将data里提供的天气数据绑定到页面里，代码如下：

```
<view class="content">
    <view class="today">
        <view class="info">
            <view class="temp">{{temp}}℃</view>
            <view class="lowhigh">{{low}}/{{high}}</view>
            <view class="type">{{type}}</view>
            <view class="city">{{city}}</view>
            <view class="week">{{week}}</view>
            <view class="weather"> {{weather}} </view>
        </view>
    </view>
</view>
```

界面效果如图2.21所示。

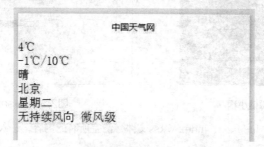

图2.21 天气界面

（6）进入index.wxss，为index.wxml添加样式，美化页面，代码如下：

```
.content{
    font-family：微软雅黑,宋体;
    font-size: 14px;
    background-size:cover;
    height: 100%;
    width:100%;
    color:#333333;
}
.today{
    padding-top:70rpx;
    height:50%;
}
.temp{
    font-size:80px;
    text-align: center;
}
.city{
    font-size: 20px;
    text-align: center;
    margin-top:20rpx;
    margin-right: 10rpx;
}
.lowhigh{
    font-size: 12px;
```

```
        text-align: center;
        margin-top: 30rpx;
}
.type{
        font-size: 16px;
        text-align: center;
        margin-top: 30rpx;
}
.week{
        font-size: 12px;
        text-align: center;
        margin-top: 30rpx;
}

.weather{
        font-size: 12px;
        text-align: center;
        margin-top: 20rpx;
}
```

添加样式后界面效果如图2.22所示。

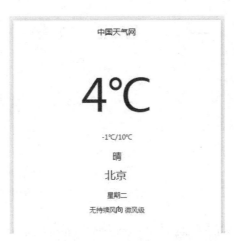

图2.22 添加样式

将js文件里的data进行数据绑定，就可以在wxml文件里通过双大括号的方式，将数据值显示出来，动态地加载数据，以实现数据绑定的动态效果。

微信小程序条件渲染

2.5 微信小程序条件渲染

2.5.1 wx:if判断单个组件

在微信小程序框架里，使用 wx:if="{{condition}}" 来判断是否需要渲染该代码块，示例代码如下：

```
<view wx:if="{{condition}}"> True </view>
```

使用 wx:elif 和 wx:else 来添加一个 else 块：

```
<view wx:if="{{length > 5}}"> 1 </view>
<view wx:elif="{{length > 2}}"> 2 </view>
<view wx:else> 3 </view>
```

2.5.2 block wx:if 判断多个组件

wx:if 是一个控制属性，需要将它添加到一个标签上。但是如果我们想一次性判断多个组件标签，就可以使用一个 <block/> 标签将多个组件包装起来，并在上面使用 wx:if 控制属性，示例代码如下：

```
<block wx:if="{{true}}">
  <view> view1 </view>
  <view> view2 </view>
</block>
```

<block/> 不是一个组件，它仅仅是一个包装元素，不会在页面中做任何渲染，只接受控制属性。

2.6 微信小程序列表渲染

微信小程序列表渲染

2.6.1 wx:for 列表渲染单个组件

在组件上使用wx:for控制属性绑定一个数组，即可使用数组中各项的数据重复渲染该组件。数组当前项的下标变量名默认为index，数组当前项的变量名默认为item，示例代码如下：

```
<view wx:for="{{array}}">
  {{index}}: {{item.message}}
</view>

Page({
  data: {
    array: [{
      message: 'foo',
    }, {
      message: 'bar'
    }]
  }
})
```

使用 wx:for-item 可以指定数组当前元素的变量名，使用 wx:for-index 可以指定数组当前下标的变量名，示例代码如下：

```
<view wx:for="{{array}}" wx:for-index="idx" wx:for-item="itemName">
  {{idx}}: {{itemName.message}}
</view>
```

2.6.2 block wx:for 列表渲染多个组件

wx:for应用在某一个组件上，如果想渲染一个包含多节点的结构块，这时wx:for需要应用在<block/>标签上，示例代码如下：

```
<block wx:for="{{[1, 2, 3]}}">
  <view> {{index}}: </view>
  <view> {{item}} </view>
</block>
```

2.6.3 wx:key 指定唯一标识符

如果列表中项目的位置会动态改变或者有新的项目添加到列表中，并且希望列表中的项目保持自己的特征和状态（如 <input/> 中的输入内容，<switch/> 的选中状态），需要使用 wx:key 来指定列表中项目的唯一标识符。wx:key 的值以以下两种形式提供。

（1）字符串：代表在 for 循环的 array 中 item 的某个 property，该 property 的值需要是列表中唯一的字符串或数字，且不能动态改变。

（2）保留关键字：*this 代表在 for 循环中的 item 本身，这种表示需要 item 本身是一个唯一的字符串或者数字，当数据改变触发渲染层重新渲染的时候，会校正带有 key 的组件，框架会确保它们被重新排序，而不是重新创建，以确保组件保持自身的状态，并且提高列表渲染时的效率。

示例代码如下：

```
<switch wx:for="{{objectArray}}" wx:key="unique" style="display: block;"> {{item.id}} </switch>
Page({
  data: {
    objectArray: [
      {id: 5, unique: 'unique_5'},
      {id: 4, unique: 'unique_4'},
      {id: 3, unique: 'unique_3'},
      {id: 2, unique: 'unique_2'},
      {id: 1, unique: 'unique_1'},
      {id: 0, unique: 'unique_0'},
    ]
  }
})
```

如不提供 wx:key，会报一个 warning，如果明确知道该列表是静态的，或者不必关注其顺序，可以选择忽略。

2.7 微信小程序定义模板

WXML提供模板（template）功能，可以把一些共用的、复用的代码，在模板中定义代码片段，然后在不同的地方调用，以达到一次编写，多次直接使用的效果。

微信小程序定义模板

2.7.1 定义模板

在<template/>内定义代码片段，使用name属性，作为模板的名字，示例代码如下：

```
<template name="msgItem">
  <view>
    <text> {{index}}: {{msg}} </text>
    <text> Time: {{time}} </text>
  </view>
</template>
```

2.7.2 使用模板

在WXML文件里，使用 is 属性，声明需要使用的模板，然后将模板所需要的 data 传入，示例代码如下：

```
<template is="msgItem" data="{{item}}"/>
Page({
  data: {
    item: {
      index: 0,
      msg: 'this is a template',
      time: '2018-06-13'
    }
```

```
    }
})
```
is 属性可以使用三元运算语法,来动态决定具体需要渲染哪个模板:
```
<template name="odd">
    <view> odd </view>
</template>
<template name="even">
    <view> even </view>
</template>

<block wx:for="{{[1, 2, 3, 4, 5]}}">
    <template is="{{item % 2 == 0 ? 'even' : 'odd'}}"/>
</block>
```

2.8 微信小程序的引用功能

微信小程序的引用功能

WXML 提供两种文件引用方式:import和include。两者的区别在于:import引用模板文件,include将引用整个除了<template/>文件。

2.8.1 import引用

import可以在该文件中使用目标文件定义的template。

假如在 item.wxml 中定义了一个叫item的template,示例代码如下:
```
<!-- item.wxml -->
<template name="item">
    <text>{{text}}</text>
</template>
```
在 index.wxml 中引用了 item.wxml,就可以使用item模板,示例代码如下:
```
<import src="item.wxml"/>
<template is="item" data="{{text: 'forbar'}}"/>
```

2.8.2 include引用

include可以将目标文件除了<template/>的整个代码引入,相当于是复制到include位置,示例代码如下:
```
<!-- index.wxml -->
<include src="header.wxml"/>
<view> body </view>
<include src="footer.wxml"/>

<!-- header.wxml -->
<view> header </view>

<!-- footer.wxml -->
<view> footer </view>
```

2.9 WXS小程序脚本语言

WXS小程序脚本语言

WXS(WeiXin Script)是小程序的一套脚本语言,结合 WXML页面文件,可以构建出页面的结构。它是把原来放在js文件里进行处理的逻辑,直接放在WXML页面文件里进行处理。它有两种使用方式:一种是将WXS脚本语言嵌入到WXML页面文件里,在wxml文件中的 <wxs> 标签内来处理相关逻辑;另一种是以.wxs后缀结尾的文件独立存在,然后再引入到WXML页面文件里使用。

第一种方式：

```
<!--wxml-->
<wxs module="m1">
var msg = "hello world";

module.exports.message = msg;
</wxs>

<view> {{m1.message}} </view>
```

第二种方式：

在指定的项目目录上单击鼠标右键可以创建.wxs文件，如图2.23所示。

图2.23　创建.wxs脚本语言文件

```
// /pages/tools.wxs
var foo = "'hello world' from tools.wxs";
var bar = function (d) {
  return d;
}
module.exports = {
  FOO: foo,
  bar: bar,
};
module.exports.msg = "some msg";
<!-- page/index/index.wxml -->
<wxs src="./../tools.wxs" module="tools" />
<view> {{tools.msg}} </view>
<view> {{tools.bar(tools.FOO)}} </view>
```

2.9.1　模块化

WXS小程序脚本语言是以模块化的形式存在的，是一个单独的模块。在一个模块里面定义的变量与函数，默认为私有的，对其他模块不可见；一个模块要想对外暴露其内部的私有变量与函数，只能通过

module.exports 实现。

```
// /pages/comm.wxs
var foo = "'hello world' from comm.wxs";
var bar = function(d) {
  return d;
}
module.exports = {
  foo: foo,
  bar: bar
};
```

在.wxs模块中引用其他 wxs 文件模块,可以使用 require 函数。在wxs文件里只能引用 .wxs 文件模块,且必须使用相对路径。wxs 模块均为单例,多个页面、多个地方、多次引用,使用的都是同一个 wxs 模块对象。如果一个 wxs 模块在定义之后,一直没有被引用,则该模块不会被解析与运行。

```
// /pages/tools.wxs
var foo = "'hello world' from tools.wxs";
var bar = function (d) {
  return d;
}
module.exports = {
  FOO: foo,
  bar: bar,
};
module.exports.msg = "some msg";

// /pages/logic.wxs
var tools = require("./tools.wxs");
console.log(tools.FOO);
console.log(tools.bar("logic.wxs"));
console.log(tools.msg);

<!-- /page/index/index.wxml -->
<wxs src="./../logic.wxs" module="logic" />
```

2.9.2 变量与数据类型

1. 变量使用

WXS 中的变量均为值的引用,如果只声明变量而不赋值,则默认值为 undefined。

```
var foo = 1;
var bar = "hello world";
var i; // i === undefined
```

变量名命名规则:

(1)首字符必须是字母(a-z,A-Z)、下画线(_);

(2)剩余字符可以是字母(a-z,A-Z)、下画线(_)、数字(0-9);

(3)保留标识符不能作为变量名,如delete、void、typeof、null、undefined、NaN、Infinity、var、if、else、true、false、require、this、function、arguments、return、for、while、do、break、continue、switch、case、default。

2. 数据类型

WXS支持的数据类型有:number数值类型、string字符串类型、boolean布尔值类型、object对象类型、function函数类型、array 数组类型、date日期类型、regexp正则类型。

（1）number数值类型。

number 包括两种数值：整数、小数。

```
var a = 10;
var PI = 3.141592653589793;
```

（2）string字符串类型。

string 有两种写法：

```
'hello world';
"hello world";
```

（3）boolean布尔值类型。

布尔值只有两个特定的值：true 和 false。

（4）object对象类型。

object 是一种无序的键值对。使用方法如下所示：

```
var o = {} //生成一个新的空对象
//生成一个新的非空对象
o = {
  'string'  : 1,   //object 的 key 可以是字符串
  const_var : 2,   //object 的 key 也可以是符合变量定义规则的标识符
  func      : {}, //object 的 value 可以是任何类型
};
//对象属性的读操作
console.log(1 === o['string']);
console.log(2 === o.const_var);

//对象属性的写操作
o['string']++;
o['string'] += 10;
o.const_var++;
o.const_var += 10;

//对象属性的读操作
console.log(12 === o['string']);
console.log(13 === o.const_var);
```

（5）function函数类型。

function 支持以下的定义方式：

```
//方法 1
function a (x) {
  return x;
}

//方法 2
var b = function (x) {
  return x;
}
```

function 同时也支持以下的语法（匿名函数、闭包等）：

```
var a = function (x) {
  return function () { return x; }
}
```

（6）array数组类型。

array 支持以下的定义方式：

```
var a = [];          //生成一个新的空数组
a = [1,"2",{},function(){}];   //生成一个新的非空数组,数组元素可以是任何类型
```

（7）date日期类型。

生成 date 对象需要使用 getDate函数，返回一个当前时间的对象。

```
getDate()
getDate(milliseconds)
getDate(datestring)
getDate(year, month[, date[, hours[, minutes[, seconds[, milliseconds]]]]])
```

参数如下。

milliseconds：从1970年1月1日00:00:00 UTC开始计算的毫秒数。

datestring：日期字符串，其格式为："month day, year hours:minutes:seconds"。

（8）regexp正则类型。

生成 regexp 对象需要使用getRegExp函数。

```
getRegExp(pattern[, flags])
```

参数如下。

pattern：正则表达式的内容。

flags：修饰符。该字段只能包含以下字符：

g: global;

i: ignoreCase;

m: multiline。

2.9.3 注释

WXS注释有3种方式：单行注释、多行注释、结尾注释。

```
<wxs module="sample">
// 方法一：单行注释
//var name = "小刚";

//方法二：多行注释
/*
var a = 1;
var b = 2;
*/

//方法三：结尾注释。即从 /* 开始往后的所有 WXS 代码均被注释
/*
var a = 1;
var b = 2;
var c = "fake";

</wxs>
```

2.9.4 语句

WXS里，可以使用if条件语句、switch条件语句、for循环语句和while循环语句。

1. if条件语句

在 WXS 中，可以使用以下格式的 if 语句：if … else if … else statement$N$。通过该句型，可以在statement1~statement$N$之间选其中一个执行。

示例语法：
```
if (表达式) {
  代码块;
} else if (表达式) {
  代码块;
} else if (表达式) {
  代码块;
} else {
  代码块;
}

var age = 10;
if(age < 18){
  console.log("未成年");
}else if(age < 28){
  console.log("青年");
}else{
  console.log("壮年");
}
```

2. switch条件语句

switch语句根据表达式的值与case变量值做比较，哪个case变量值与表达式值相等就执行哪个case语句。default 分支可以省略不写，case 关键词后面只能使用变量、数字或字符串。如果不写break结束语句，程序就会向下继续执行其他满足条件的case语句。

示例语法：
```
switch (表达式) {
  case 变量:
    语句;
  case 数字:
    语句;
    break;
  case 字符串:
    语句;
  default:
    语句;
}

var exp = 10;
switch ( exp ) {
case "10":
  console.log("string 10");
  break;
case 10:
  console.log("number 10");
  break;
case exp:
  console.log("var exp");
  break;
default:
  console.log("default");
}
```

3. for循环语句

for循环语句，用来遍历集合，支持使用 break、continue 关键词。

示例语法：

```
for (语句; 语句; 语句) {
    代码块;
}
for (var i = 0; i < 3; ++i) {
    console.log(i);
    if( i >= 1) break;
}
```

4. while 循环语句

示例语法：

```
while (表达式){
    代码块;
}

do {
    代码块;
} while (表达式)
```

当表达式为 true 时，循环执行语句或代码块。

支持使用 break、continue 关键词。

2.10 沙场大练兵：仿香哈菜谱微信小程序

仿香哈菜谱微信小程序

香哈菜谱是围绕美食而做的一款小程序，在这里可以查看各式各样的菜谱。对于菜谱类App软件，用户使用的频率不高。当碰到不会做的菜式或者想做一些新的菜式时，用户才会去App软件查看，而微信小程序就可以满足这种低频率使用的场景，如图2.24、图2.25所示。

图2.24 学做菜

图2.25 头条

2.10.1 底部标签导航设计

仿香哈菜谱微信小程序，底部有5个导航标签：学做菜、头条、美食圈、消息、我的。标签导航选中时，导航图标会变为红色，导航文字也会变为红色，如图2.26所示。

（1）新建一个香哈菜谱xhcp项目，如图2.27所示。

图2.26　底部标签导航选中效果　　　　　图2.27　添加项目

（2）将准备好的底部标签导航图标、美食轮播图片、宫格导航图标、香哈头条美食图片复制到pages文件夹下。

（3）打开app.json配置文件，在pages数组里添加5个页面路径"pages/cook/cook""pages/headline/headline""pages/food/food""pages/message/message""pages/me/me"，保存后会自动生成相应的页面文件夹；删除"pages/index/index""pages/logs/logs"页面路径以及对应的文件夹，如图2.28所示。

图2.28　配置页面路径

（4）在window数组里配置窗口导航背景颜色为灰色（#494949），导航栏文字为"学做菜"，字体颜色设置为白色（#ffffff），具体配置如图2.29所示。

```
app.json ×
1  {
2      "pages":[
3          "pages/cook/cook",
4          "pages/headline/headline",
5          "pages/food/food",
6          "pages/message/message",
7          "pages/me/me"
8      ],
9      "window":{
10         "backgroundTextStyle": "light",
11         "navigationBarBackgroundColor": "#494949",
12         "navigationBarTitleText": "学做菜",
13         "navigationBarTextStyle": "#ffffff"
14     }
15 }
```

图2.29 窗口及导航栏配置

（5）在tabBar对象里配置底部标签导航背景色为白色（#ffffff），文字默认颜色为灰色（#999999），选中时为红色（#CC1004），在list数组里配置底部标签导航对应的页面、导航名称、默认时图标、选中时图标，具体配置如图2.30所示。

这样就完成了仿菜谱微信小程序的底部标签导航配置，单击不同的导航标签，可以切换显示不同的页面，同时导航图标和导航文字会呈现为选中状态，如图2.31所示。

```
app.json ×
15     "tabBan": {
16         "backgroundColor":"#ffffff",
17         "color":"#999999",
18         "selectedColor":"#CC1004",
19         "borderStyle":"black",
20         "list": [{
21             "pagePath": "pages/cook/cook",
22             "text": "学做菜",
23             "iconPath": "pages/images/tab/cook-0.jpg",
24             "selectedIconPath": "pages/images/tab/cook-1.jpg"
25         },{
26             "pagePath": "pages/headline/headline",
27             "text": "头条",
28             "iconPath": "pages/images/tab/headline-0.jpg",
29             "selectedIconPath": "pages/images/tab/headline-1.jpg"
30         },{
31             "pagePath": "pages/food/food",
32             "text": "美食圈",
33             "iconPath": "pages/images/tab/food-0.jpg",
34             "selectedIconPath": "pages/images/tab/food-0.jpg"
35         },{
36             "pagePath": "pages/message/message",
37             "text": "消息",
38             "iconPath": "pages/images/tab/message-0.jpg",
39             "selectedIconPath": "pages/images/tab/message-1.jpg"
40         },{
41             "pagePath": "pages/me/me",
42             "text": "我的",
43             "iconPath": "pages/images/tab/me-0.jpg",
44             "selectedIconPath": "pages/images/tab/me-1.jpg"
45         }]
46     }
47 }
```

图2.30 底部标签导航配置

图2.31 学做菜界面

2.10.2 宫格导航设计

在学做菜这个界面里，有海报轮播的图片（在微信小程序里有专门的swiper滑块视图组件实现这个效果，在3.1.3节中会详细讲解），还有4个宫格导航：菜谱分类、视频、美食养生和闪购，如图2.32所示。

图2.32 海报图片和宫格导航

（1）进入到pages/cook/cook.wxml文件，先设计海报轮播区域，使用一张图片来进行布局，图片的宽度设置为100%，高度设置为230px，具体代码如下。

```
<view class="content">
  <view class="img">
    <image src="../images/haibao/haibao-1.jpg" style="width:100%;height:230px;"></image>
  </view>
</view>
```

界面效果如图2.33所示。

图2.33 海报图片

（2）设计宫格导航，分为4个导航：菜谱分类、视频、美食养生、闪购。每个导航对应一个图标，在导航的下面是灰色的间隔线，具体代码如下。

```html
<view class="content">
  <view class="img">
    <image src="../images/haibao/haibao-1.jpg" style="width:100%;height:230px;"></image>
  </view>

  <view class="nav">
    <view class="nav-item">
      <view> <image src="../images/icon/fenlei.jpg" style="width:25px;height:23px;"></image></view>
      <view>菜谱分类</view>
    </view>
    <view class="nav-item">
      <view> <image src="../images/icon/shipin.jpg" style="width:25px;height:23px;"></image></view>
      <view>视频</view>
    </view>
    <view class="nav-item">
      <view> <image src="../images/icon/meishi.jpg" style="width:25px;height:23px;"></image></view>
      <view>美食养生</view>
    </view>
    <view class="nav-item">
      <view> <image src="../images/icon/shangou.jpg" style="width:25px;height:23px;"></image></view>
      <view>闪购</view>
    </view>
  </view>
  <view class="hr"></view>
</view>
```

（3）进入到pages/cook/cook.wxss文件里，针对宫格导航，添加样式，具体代码如下。

```css
.nav{
  display: flex;
  flex-direction: row;
  text-align: center;
}
.nav-item{
  width: 25%;
  margin-top:20px;
  font-size: 12px;
}
.hr{
  height: 15px;
  background-color: #cccccc;
  margin-top:15px;
  opacity: 0.2;
}
.head{
  display: flex;
  flex-direction: row;
  margin: 10px;
  font-size: 13px;
  color: #999999;
}
```

```
.right{
    position: absolute;
    right: 10px;
    color: #cccccc;
}
.hr2{
    height: 2px;
    background-color: #cccccc;
    opacity: 0.2;
}
```

界面效果如图2.34所示。

图2.34 宫格导航

这样就完成了海报轮播区域和宫格导航区域的界面布局，在很多App上都会用海报轮播和宫格导航的方式来进行界面布局。

2.10.3 香哈头条初始化数据

微信小程序作为客户端，它的数据来源于服务端。下面模拟一下服务端提供的香哈头条列表的数据，有了数据，页面才能动态地进行渲染。

进入到pages/cook/cook.js文件里，添加initData函数，在data页面初始化数据里添加array数组，然后将initData定义的数据通过setData设值函数赋值给array数组，具体代码如下。

```
Page({
  data:{
    array:[]
  },
```

```
onLoad:function(options){
    var array = this.initData();
    this.setData({array:array});
},
initData:function(){
    var array = [];
    var object1 = new Object();
    object1.img = '../images/list/food-1.jpg';
    object1.title='爱心早餐';
    object1.type='健康养生';
    object1.liulan='20696浏览';
    object1.pinglun='7评论';
    array[0] = object1;

    var object2 = new Object();
    object2.img = '../images/list/food-2.jpg';
    object2.title='困了只想喝咖啡';
    object2.type='家庭医生在线';
    object2.liulan='29628浏览';
    object2.pinglun='13评论';
    array[1] = object2;

    var object3 = new Object();
    object3.img = '../images/list/food-3.jpg';
    object3.title='橘子吃多变小黄人';
    object3.type='家庭医生在线';
    object3.liulan='19585浏览';
    object3.pinglun='6评论';
    array[2] = object3;

    var object4 = new Object();
    object4.img = '../images/list/food-4.jpg';
    object4.title='搜狐新闻，手机用久了';
    object4.type='广告';
    object4.liulan='4688浏览';
    object4.pinglun='4评论';
    array[3] = object4;

    var object5 = new Object();
    object5.img = '../images/list/food-5.jpg';
    object5.title='困了只想喝咖啡';
    object5.type='家庭医生在线';
    object5.liulan='29628浏览';
    object5.pinglun='13评论';
    array[4] = object5;

    return array;
}
```

2.10.4 香哈头条列表渲染及绑定数据

香哈头条里有菜谱的图片、美食名称、分类、浏览数量以及评论数量，如图2.35所示。

图2.35　香哈头条列表

（1）进入到pages/cook/cook.wxml文件里，进行香哈头条列表信息界面布局，具体代码如下。

```
<view class="content">
  <view class="img">
    <image src="../images/haibao/haibao-1.jpg" style="width:100%;height:230px;"></image>
  </view>

  <view class="nav">
    <view class="nav-item">
        <view> <image src="../images/icon/fenlei.jpg" style="width:25px;height:23px;"></image></view>
        <view>菜谱分类</view>
    </view>
    <view class="nav-item">
        <view> <image src="../images/icon/shipin.jpg" style="width:25px;height:23px;"></image></view>
        <view>视频</view>
    </view>
    <view class="nav-item">
        <view> <image src="../images/icon/meishi.jpg" style="width:25px;height:23px;"></image></view>
        <view>美食养生</view>
    </view>
    <view class="nav-item">
        <view> <image src="../images/icon/shangou.jpg" style="width:25px;height:23px;"></image></view>
        <view>闪购</view>
    </view>
  </view>
  <view class="hr"></view>
  <view class="head">
```

```
          <view>香哈头条</view>
          <view class="right"></view>
      </view>
      <view class="list">
          <view class="item" bindtap="seeDetail" id="0">
              <view>
                  <image src="../images/list/food-1.jpg" style="width:75px;height:58px;"></image>
              </view>
              <view class="desc">
                  <view class="title">爱心早餐</view>
                  <view class="count">
                      <view>健康养生</view>
                      <view class="liulan">20696浏览</view>
                      <view class="pinglun">7评论</view>
                  </view>
              </view>
          </view>
          <view class="hr2"></view>
      </view>
  </view>
```

（2）进入到pages/cook/cook.wxss文件里，针对香哈头条列表信息添加样式，具体代码如下。

```
.item{
    display: flex;
    flex-direction: row;
    margin-left: 10px;
    margin-bottom:5px;
}
.desc{
    margin-left: 20px;
    line-height: 30px;
}
.title{
    font-weight: bold;
}
.count{
    display: flex;
    flex-direction: row;
    font-size: 12px;
    color: #999999;
}
.liulan{
    position: absolute;
    right: 70px;
}
.pinglun{
    position: absolute;
    right: 10px;
}
```

界面效果如图2.36所示。

（3）现在香哈头条列表数据直接写在界面里，下面通过数据绑定和wx:for循环的方式动态加载数据，具体代码如下。

图2.36 香哈头条列表界面效果

```
<view class="content">
  <view class="img">
    <image src="../images/haibao/haibao-1.jpg" style="width:100%;height:230px;"></image>
  </view>

  <view class="nav">
    <view class="nav-item">
      <view> <image src="../images/icon/fenlei.jpg" style="width:25px;height:23px;"></image></view>
      <view>菜谱分类</view>
    </view>
    <view class="nav-item">
      <view> <image src="../images/icon/shipin.jpg" style="width:25px;height:23px;"></image></view>
      <view>视频</view>
    </view>
    <view class="nav-item">
      <view> <image src="../images/icon/meishi.jpg" style="width:25px;height:23px;"></image></view>
      <view>美食养生</view>
    </view>
    <view class="nav-item">
      <view> <image src="../images/icon/shangou.jpg" style="width:25px;height:23px;"></image></view>
      <view>闪购</view>
    </view>
  </view>
  <view class="hr"></view>
  <view class="head">
    <view>香哈头条</view>
    <view class="right"></view>
  </view>
```

```
<view class="list">
  <block wx:for="{{array}}">
    <view class="item" bindtap="seeDetail" id="0">
      <view>
        <image src="{{item.img}}" style="width:75px;height:58px;"></image>
      </view>
      <view class="desc">
        <view class="title">{{item.title}}</view>
        <view class="count">
          <view>{{item.type}}</view>
          <view class="liulan">{{item.liulan}}</view>
          <view class="pinglun">{{item.pinglun}}</view>
        </view>
      </view>
    </view>
    <view class="hr2"></view>
  </block>
</view>
</view>
```

界面效果如图2.37所示。

图2.37 香哈头条列表展现

在完成香哈头条列表展现时，首先进行列表的界面布局，可以把数据写在页面里。页面布局完成后通过数据绑定的方式，将js里的data数据和WXML界面进行绑定，达到一种动态获取数据的效果。

2.10.5 香哈头条模板引用

很多页面都采用香哈头条列表信息这样的展现方式，因此就可以把它制作成模板，达到一次制作、多次使用的效果。

（1）在pages添加一个template目录，再添加一个template.wxml文件，在这个文件里制作一个香哈头条列表的模板，模板名称为cooks，将列表循环的内容放置在这个文件里，如图2.38所示。

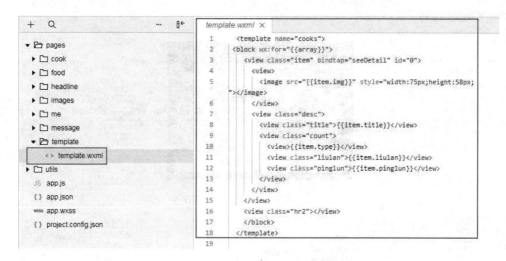

图2.38　香哈头条列表模板

（2）将cooks模板引入到cook.wxml里，将列表信息展现出来，具体代码如下。

```
<view class="content">
  <view class="img">
    <image src="../images/haibao/haibao-1.jpg" style="width:100%;height:230px;"></image>
  </view>

  <view class="nav">
    <view class="nav-item">
        <view> <image src="../images/icon/fenlei.jpg" style="width:25px;height:23px;"></image></view>
        <view>菜谱分类</view>
    </view>
    <view class="nav-item">
        <view> <image src="../images/icon/shipin.jpg" style="width:25px;height:23px;"></image></view>
        <view>视频</view>
    </view>
    <view class="nav-item">
        <view> <image src="../images/icon/meishi.jpg" style="width:25px;height:23px;"></image></view>
        <view>美食养生</view>
    </view>
    <view class="nav-item">
        <view> <image src="../images/icon/shangou.jpg" style="width:25px;height:23px;"></image></view>
        <view>闪购</view>
    </view>
  </view>
  <view class="hr"></view>
  <view class="head">
     <view>香哈头条</view>
     <view class="right">></view>
  </view>
<import src="../template/template" />
<view class="list">
```

```
        <template is="cooks" data="{{array}}"/>
    </view>
</view>
```

界面效果如图2.39所示。

图2.39 香哈头条模板使用

可以看出使用模板前后界面效果是一样的，模板可以被不同的地方使用，像香哈头条列表一样，在其他的地方也有列表的展现，这时就可以使用模板的方式，达到一次编写、多次引用的效果。

2.11 小结

本章讲解了微信小程序的框架，涉及很多内容，应重点掌握如下内容：

（1）了解微信小程序目录结构，理解框架全局文件、工具类文件、框架页面文件的使用；

（2）会配置窗口导航栏以及底部标签导航；

（3）了解微信小程序注册程序应用以及生命周期函数的意义和使用；

（4）掌握微信小程序注册页面的使用，包括页面初始化数据、生命周期函数的使用、页面相关事件处理函数的使用、页面路由管理和setData设置函数的使用；

（5）学会微信小程序如何绑定数据；

（6）学会微信小程序条件判断和列表渲染的使用；

（7）学会微信小程序模板的定义和引用。

第3章
用微信小程序组件构建UI界面

- 视图容器组件
- 基础内容组件
- 丰富的表单组件
- 导航组件
- 媒体组件
- 地图组件
- 画布组件
- 沙场大练兵：表单登录注册微信小程序
- 小结

微信小程序框架里提供了很多UI组件，这些UI组件就像积木一样。使用积木可以搭建一座房子、一座大桥，我们使用UI组件来搭建小程序界面。每个组件都有不同的用处，比如，用来包裹内容的视图容器组件、用来呈现内容的基础内容组件、丰富的表单组件、页面链接的导航组件、视频音频播放的媒体组件、地图组件和画布组件。有了这些组件，就可以完成界面的布局和界面的渲染。

3.1 视图容器组件

视图容器组件

视图容器组件共有5种：view视图容器、scroll-view可滚动视图区域、swiper滑块视图容器、movable-view可移动视图容器、cover-view覆盖原生组件的视图容器。

3.1.1 view视图容器

view视图容器是WXML界面布局的基础组件，它的使用和HTML里的DIV类似，主要用于界面的布局。view视图容器也有自己的属性，如表3.1所示。

表3.1 view属性

属性	类型	默认值	说明
hover	Boolean	false	是否启用单击态
hover-class	String	none	指定按下去的样式类。当 hover-class="none"时，没有单击态效果
hover-start-time	Number	50	按住后多久出现单击态，单位毫秒
hover-stay-time	Number	400	手指松开后单击态保留时间，单位毫秒

在WXML界面里使用view布局，渲染出界面内容，如图3.1所示。

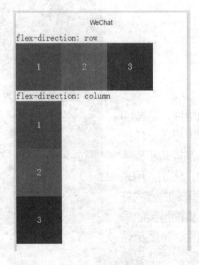

图3.1 view布局

具体代码如下：

```
<view class="section">
    <view class="section__title">flex-direction: row</view>
    <view class="flex-wrp" style="display:flex;flex-direction:row;">
        <view class="flex-item bc_green" style="width:100px;height:100px;background-color:green;color:#ffffff;text-align:center;line-height:100px;">1</view>
        <view class="flex-item bc_red" style="width:100px;height:100px;background-color:red;color:#ffffff;text-align:center;line-height:100px;">2</view>
        <view class="flex-item bc_blue" style="width:100px;height:100px;background-color:blue;color:#ffffff;text-align:center;line-height:100px;">3</view>
    </view>
</view>
```

```
<view class="section">
  <view class="section__title">flex-direction: column</view>
  <view class="flex-wrp" style="display:flex;height: 300px;flex-direction:column;">
    <view class="flex-item bc_green" style="width:100px;height:100px;background-color:green;color:#ffffff;text-align:center;line-height:100px;">1</view>
    <view class="flex-item bc_red" style="width:100px;height:100px;background-color:red;color:#ffffff;text-align:center;line-height:100px;">2</view>
    <view class="flex-item bc_blue" style="width:100px;height:100px;background-color:blue;color:#ffffff;text-align:center;line-height:100px;">3</view>
  </view>
</view>
```

3.1.2 scroll-view可滚动视图区域

scroll-view可滚动视图区域允许视图区域内容横向滚动或者纵向滚动，类似于浏览器的横向滚动条和垂直滚动条，scroll-view拥有自己的属性和事件，如表3.2所示。

表3.2　scroll-view属性

属性	类型	默认值	说明
scroll-x	Boolean	false	允许横向滚动
scroll-y	Boolean	false	允许纵向滚动
upper-threshold	Number	50	距顶部/左边多远时（单位px），触发 scrolltoupper 事件
lower-threshold	Number	50	距底部/右边多远时（单位px），触发 scrolltolower 事件
scroll-top	Number		设置竖向滚动条位置
scroll-left	Number		设置横向滚动条位置
scroll-into-view	String		值应为某子元素id，则滚动到该元素，元素顶部对齐滚动区域顶部
bindscrolltoupper	EventHandle		滚动到顶部/左边，会触发 scrolltoupper 事件
bindscrolltolower	EventHandle		滚动到底部/右边，会触发 scrolltolower 事件
bindscroll	EventHandle		滚动时触发，event.detail = {scrollLeft, scrollTop, scrollHeight, scrollWidth, deltaX, deltaY}

1．纵向滚动

允许内容纵向滚动，需要给<scroll-view/>一个固定高度，可以绑定滚动到顶部/左边（bindscrolltoupper）、滚动到底部/右边（bindscrolltolower）、滚动时（bindscroll）触发的事件，也可以滚动到指定的id区域（scroll-into-view）。下面实现纵向滚动，如图3.2所示。

图3.2　纵向滚动

（1）在wxml文件里使用scroll-view进行布局，设置scroll-y="true"纵向滚动，绑定bindscrolltoupper、bindscrolltolower、bindscroll、scroll-into-view、scroll-top事件，具体代码如下。

```
<view class="section">
  <view class="section__title">scroll-view纵向滚动</view>
  <scroll-view scroll-y="true" style="height: 200px;" bindscrolltoupper="upper" bindscrolltolower="lower" bindscroll="scroll" scroll-into-view="{{toView}}" scroll-top="{{scrollTop}}">
    <view id="green"  style="width:100%;height:100px;background-color:green;"></view>
    <view id="red"    style="width:100%;height:100px;background-color:red;"></view>
    <view id="yellow" style="width:100%;height:100px;background-color:yellow;"></view>
    <view id="blue"   style="width:100%;height:100px;background-color:blue;"></view>
  </scroll-view>

  <view class="btn-area">
    <button type="default" style="margin:10px;" bindtap="tap">click me to scroll into view </button>
    <button type="default" style="margin:10px;"  bindtap="tapMove">click me to scroll</button>
  </view>
```

（2）在js文件里设置颜色的数组，绑定toView和scrollTop数据值，提供bindscrolltoupper、bindscrolltolower、bindscroll、scroll-into-view、scroll-top事件函数，具体代码如下。

```
var order = ['red', 'yellow', 'blue', 'green', 'red']
Page({
  data: {
    toView: 'red',
    scrollTop: 100
  },
  upper: function(e) {
    console.log(e)
  },
  lower: function(e) {
    console.log(e)
  },
  scroll: function(e) {
    console.log(e)
  },
  tap: function(e) {
    for (var i = 0; i < order.length; ++i) {
      if (order[i] === this.data.toView) {
        this.setData({
          toView: order[i + 1]
        })
        break
      }
    }
  },
  tapMove: function(e) {
    this.setData({
      scrollTop: this.data.scrollTop + 10
    })
  }
})
```

这样就可以实现纵向滚动了，可以滚动到指定区域，也可以滚动到指定的位置，同时滚动到顶部或底部会触发相应的事件，在滚动过程中也可以触发相应的事件。

2. 横向滚动

在使用"今日头条"或"腾讯新闻"时，在新闻列表的上方都会有新闻频道供我们选择，可以向左滑动和向右滑动来查看相应类别的新闻，可以采用scroll-view来实现这些新闻频道的横向滚动，如图3.3所示。

图3.3 今日头条新闻频道

在wxml文件里使用scroll-view进行布局，设置scroll-x="true"横向滚动，具体代码如下。

```
<view class="section">
  <view class="section__title">新闻频道横向滚动</view>
  <scroll-view scroll-x="true" style="width: 100%;">
    <view style="display:flex;flex-direction:row">
      <view style="margin-right:10px;">推荐</view>
      <view style="margin-right:10px;">视频</view>
      <view style="margin-right:10px;">热点</view>
      <view style="margin-right:10px;">本地</view>
      <view style="margin-right:10px;">社会</view>
      <view style="margin-right:10px;">娱乐</view>
      <view style="margin-right:10px;">科技</view>
      <view style="margin-right:10px;">汽车</view>
      <view style="margin-right:10px;">体育</view>
      <view style="margin-right:10px;">财经</view>
      <view style="margin-right:10px;">军事</view>
      <view style="margin-right:10px;">国际</view>
      <view style="margin-right:10px;">时尚</view>
      <view style="margin-right:10px;">游戏</view>
      <view style="margin-right:10px;">美文</view>
    </view>
  </scroll-view>
</view>
```

这样就可以实现横向滚动了，可以向左滑动和向右滑动。

（1）请勿在 scroll-view 中使用 textarea、map、canvas、video 组件。
（2）scroll-into-view 的优先级高于 scroll-top。
（3）在滚动 scroll-view 时会阻止页面回弹，所以在 scroll-view 中滚动，是无法触发 onPullDownRefresh 的。
（4）若要使用下拉刷新，请使用页面的滚动，而不是 scroll-view，这样也能通过单击顶部状态栏回到页面顶部。

3.1.3　swiper滑块视图容器

swiper滑块视图容器用来在指定区域内切换显示内容，常用来制作海报轮播效果和页签内容切换效果，它的属性如表3.3所示。

表3.3　swiper属性

属性	类型	默认值	说明
indicator-dots	Boolean	false	是否显示面板指示点
autoplay	Boolean	false	是否自动切换
current	Number	0	当前所在页面的 index
interval	Number	5000	自动切换时间间隔
duration	Number	500	滑动动画时长
circular	Boolean	false	是否采用衔接滑动
bindchange	EventHandle		current 改变时会触发 change 事件，event.detail = {current: current}

1. 海报轮播效果

海报轮播效果常用来展示商品图片信息或者广告信息，是很多网站或者App软件都会采用的一种布局方式，如图3.4、图3.5所示。

图3.4　海报1　　　　　　　　　　　图3.5　海报2

（1）在wxml文件里，采用swiper滑块视图容器组件进行海报轮播区域的布局，具体代码如下：

```
<view class="haibao">
    <swiper indicator-dots="{{indicatorDots}}" autoplay="{{autoplay}}" interval="{{interval}}" duration="{{duration}}">
        <block wx:for="{{imgUrls}}">
            <swiper-item>
                <image src="{{item}}" class="silde-image" style="width:100%"></image>
```

```
      </swiper-item>
    </block>
  </swiper>
</view>
```

（2）在js文件里，提供海报轮播的图片，设置是否自动播放，提供轮播的时长等数据，通过数据绑定的方式渲染到页面上，具体代码如下。

```
Page({
  data:{
    indicatorDots:true,
    autoplay:true,
    interval:5000,
    duration:1000,
    imgUrls:[
      "http://img06.tooopen.com/images/20160818/tooopen_sy_175866434296.jpg", "http://img06.tooopen.com/images/20160818/tooopen_sy_175833047715.jpg","http://img02.tooopen.com/images/20150928/tooopen_sy_143912755726.jpg"
    ]
  }
})
```

设置autoplay等于true时就可以自动进行海报轮播，设置indicatorDots等于true，代表显示面板指示点，同时可以设置interval自动切换时长、duration滑动动画时长。

2. 页签内容切换效果

swiper滑块视图容器除了可以用来实现海报轮播效果，还可以实现页签切换效果。页签切换效果常用于多种方式的登录或者多种类别的切换，如图3.6所示。

图3.6　页签切换效果

（1）进入到wxml文件里，进行账号密码登录和手机快捷登录的界面布局设计，具体代码如下。

```
<view class="content">
  <view class="loginTitle">
    <view class="{{currentTab==0?'select':'default'}}" data-current="0" bindtap="switchNav">账号密码登录</view>
    <view class="{{currentTab==1?'select':'default'}}" data-current="1" bindtap="switchNav">手机快捷登录</view>
  </view>
  <view class="hr"></view>
  <swiper current="{{currentTab}}" style="height:{{winHeight}}px">
    <swiper-item>
      <view style="margin-top:10px;border:1px solid #cccccc;width:99%;height:200px;">
        我是用来进行账号密码登录的区域
```

```
            </view>
        </swiper-item>
        <swiper-item>
         <view style="margin-top:10px;border:1px solid #cccccc;width:99%;height:200px;">
            我是用来进行手机快捷登录的区域
         </view>
        </swiper-item>
    </swiper>
</view>
```

（2）进入到wxss文件里，给页面文件添加样式，具体代码如下。

```
.loginTitle{
    display: flex;
    flex-direction: row;
    width: 100%;
}
.select{
    font-size:12px;
    color: red;
    width: 50%;
    text-align: center;
    height: 45px;
    line-height: 45px;
    border-bottom:5rpx solid red;
}
.default{
    font-size:12px;
    margin: 0 auto;
    padding: 15px;
}
.hr{
    border: 1px solid #cccccc;
    opacity: 0.2;
}
```

（3）进入到js文件里，提供窗口的宽度、高度、当前面板的索引值，提供页签切换函数，具体代码如下。

```
Page({
  data:{
    currentTab:0,
    winWidth:0,
    winHeight:0
  },
  onLoad:function(options){
    var page = this;
    wx.getSystemInfo({
      success: function(res) {
        console.log(res);
        page.setData({winWidth:res.windowWidth});
        page.setData({winHeight:res.windowHeight});
      }
    })
  },
```

```
switchNav:function(e){
    var page = this;
    if(this.data.currentTab == e.target.dataset.current){
        return false;
    }else{
        page.setData({currentTab:e.target.dataset.current});
    }
}
})
```

这样就可以实现在两种登录状态之间切换的效果了。页签切换时，页签的标题呈现为选中的状态，同时对应的内容也随之进行切换。

3.1.4 movable-view可移动视图容器

movable-view是一个可移动视图容器，在页面中可以做拖曳滑动。在使用这个组件的时候，需要先定义可移动区域movable-area，然后定义直接子节点movable-view，否则不能移动。movable-area 必须设置width和height属性，不设置默认为10px；movable-view 必须设置width和height属性，不设置默认为10px，movable-view 默认为绝对定位，top和left属性为0px。movable-view可移动视图容器的属性如表3.4所示。

表3.4 movable-view属性

属性	类型	默认值	说明
direction	String	none	movable-view的移动方向，属性值有all、vertical、horizontal、none
inertia	Boolean	false	movable-view是否带有惯性
out-of-bounds	Boolean	false	超过可移动区域后，movable-view是否还可以移动
x	Number / String		定义x轴方向的偏移，如果x的值不在可移动范围内，会自动移动到可移动范围；改变x的值会触发动画
y	Number / String		定义y轴方向的偏移，如果y的值不在可移动范围内，会自动移动到可移动范围；改变y的值会触发动画
damping	Number	20	阻尼系数，用于控制x或y改变时的动画和过界回弹的动画，值越大移动越快
friction	Number	2	摩擦系数，用于控制惯性滑动的动画，值越大摩擦力越大，滑动越快停止；必须大于0，否则会被设置成默认值
disabled	Boolean	false	是否禁用
scale	Boolean	false	是否支持双指缩放，默认缩放手势生效区域是在movable-view内
scale-min	Number	0.5	定义缩放倍数最小值
Scale-max	Number	10	定义缩放倍数最大值
scale-value	Number	1	定义缩放倍数，取值范围为 0.5~10
bindchange	EventHandle		拖动过程中触发的事件，event.detail = {x: x, y: y, source: source}，其中source表示产生移动的原因，值可为touch（拖动）、touch-out-of-bounds（超出移动范围）、out-of-bounds（超出移动范围后的回弹）、friction（惯性）和空字符串（setData）
bindscale	EventHandle		缩放过程中触发的事件，event.detail = {scale: scale}

movable-view提供了两个特殊事件：htouchmove事件，指初次手指触摸后的移动为横向移动，如果catch此事件，则意味着touchmove事件也被catch；vtouchmove事件，指初次手指触摸后的移动为纵向移

动,如果catch此事件,则意味着touchmove事件也被catch。

下面使用movable-view可移动视图容器组件来进行滑动,黄色区域代表可以移动的区域,红色方块代表可以移动的组件,如图3.7所示。

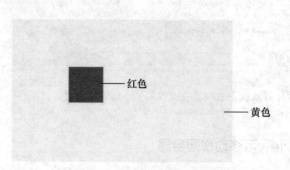

图3.7 可移动视图容器

(1)在wxml文件里,使用movable-area和movable-view视图容器组件进行布局,具体代码如下。

```
<view class="section">
  <movable-area style="height: 200px; width:100%; background: yellow;">
    <movable-view style="height: 50px; width: 50px; background: red;" x="{{x}}" y="{{y}}" direction="all">
    </movable-view>
  </movable-area>
</view>
```

(2)在js文件里,提供拖动函数、缩放函数,通过数据绑定的方式渲染到页面上,具体代码如下。

```
Page({
  data: {
    x: 0,
    y: 0
  },
  tap: function (e) {
    this.setData({
      x: 30,
      y: 30
    });
  },
  onChange: function (e) {
    console.log(e.detail)
  },
  onScale: function (e) {
    console.log(e.detail)
  }
})
```

3.1.5 cover-view覆盖原生组件的视图容器

cover-view、cover-image这两个是覆盖原生组件的视图容器。比如在使用地图组件时,地图组件本身的功能很有局限,但是想放置一些特殊的内容或图片,这时就需要使用覆盖地图组件的视图容器。

cover-view是指覆盖在原生组件之上的文本视图,可覆盖的原生组件包括map、video、canvas、camera,只支持嵌套cover-view、cover-image。

cover-image是指覆盖在原生组件之上的图片视图,可覆盖的原生组件同cover-view一样,支持嵌套cover-view。

下面使用cover-view、cover-image覆盖原生组件的视图容器组件，在video视频播放组件上放置播放、暂停两个图片，同时放置一个时间内容显示区域，如图3.8、图3.9所示。

图3.8　视频播放

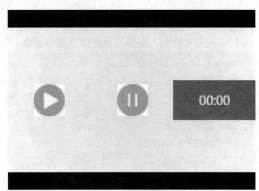

图3.9　覆盖视频播放组件

（1）在wxml文件里使用cover-view、cover-image覆盖原生组件的视图容器组件进行布局，具体代码如下。

```
<video id="myVideo" src="http://wxsnsdy.tc.qq.com/105/20210/snsdyvideodownload?filekey=30280201010
421301f0201690402534804102ca905ce620b1241b726bc41dcff44e00204012882540400&bizid=1023&hy=
SH&fileparam=302c02010104253023020204136ffd93020457e3c4ff02024ef202031e8d7f02030f42400204045a32
0a0201000400" controls="{{false}}" event-model="bubble" style="width:100%">
  <cover-view class="controls">
    <cover-view class="play" bindtap="play">
      <cover-image class="img" src="/images/play.jpg" />
    </cover-view>
    <cover-view class="pause" bindtap="pause">
      <cover-image class="img" src="/images/pause.jpg" />
    </cover-view>
    <cover-view class="time">00:00</cover-view>
  </cover-view>
</video>
```

（2）在wxss文件里添加样式，具体代码如下。

```
.controls {
  position: relative;
  top: 50%;
  height: 50px;
  margin-top: -25px;
  display: flex;
}
.play,.pause,.time {
  flex: 1;
  height: 100%;
}
.time {
  text-align: center;
  background-color: rgba(0, 0, 0, .5);
  color: white;
  line-height: 50px;
}
```

```
.img {
  width: 40px;
  height: 40px;
  margin: 5px auto;
}
```

（3）在js文件里，提供视频播放、暂停函数，初始化视频播放组件，具体代码如下。

```
Page({
  onReady() {
    this.videoCtx = wx.createVideoContext('myVideo')
  },
  play() {
    this.videoCtx.play()
  },
  pause() {
    this.videoCtx.pause()
  }
})
```

3.2 基础内容组件

基础内容组件包括icon图标组件、text文本组件、progree进度条组件、rich-text富文本组件。

基础内容组件

3.2.1 icon图标

微信小程序提供了丰富的图标组件，这些图标组件应用于不同的场景，有成功、警告、提示、取消、下载等不同含义，如图3.10所示。

图3.10　图标

icon图标组件有3个属性：图标的类型type、图标的大小size和图标的颜色color，如表3.5所示。

表3.5　图标属性

属性	类型	默认值	说明
type	String		icon的类型，有效值有：success、success_no_circle、info、warn、waiting、cancel、download、search、clear
size	Number	23	icon的大小，单位为px
color	Color		icon的颜色，同css的color

如何绘制出如图3.10所示的图标呢?

(1) 在wxml文件里,利用icon组件进行界面布局,具体代码如下。

```
<view class="group">
  <block wx:for="{{iconSize}}">
    <icon type="success" size="{{item}}"/>
  </block>
</view>

<view class="group">
  <block wx:for="{{iconType}}">
    <icon type="{{item}}" size="45"/>
  </block>
</view>

<view class="group">
  <block wx:for="{{iconColor}}">
    <icon type="success" size="45" color="{{item}}"/>
  </block>
</view>
```

(2) 在js文件里,给图标的大小、颜色和类型提供数据,具体代码如下。

```
Page({
  data: {
    iconSize: [20, 30, 40, 50, 60, 70],
    iconColor: [
      'red', 'orange', 'yellow', 'green', 'rgb(0,255,255)', 'blue', 'purple'
    ],
    iconType: [
      'success', 'info', 'warn', 'waiting', 'safe_success', 'safe_warn',
      'success_circle', 'success_no_circle', 'waiting_circle', 'circle', 'download',
      'info_circle', 'cancel', 'search', 'clear'
    ]
  }
})
```

这样就可以绘制出不同颜色、含义和大小的图标,我们根据自己的需求,利用icon组件来设计图标。

3.2.2 text文本

text文本组件支持转义符"\",比如换行\n、空格\t。<text/> 组件内只支持 <text/> 嵌套,除了文本节点以外的其他节点都无法长按选中。

下面我们来看转义符的使用,具体代码如下:

```
<view class="btn-area">
  <view class="body-view">
    <text>我爱北京\t我爱中国</text>
    <text>我爱北京\n我爱中国</text>
  </view>
</view>
```

界面效果如图3.11所示。

从图3.11中可以看出,\t具有空格功能,\n具有换行功能,同时也可以看出text文本组件是放置在同一行里,这一点不同于view组件,每个view组件都是单独一行。

图3.11 转义符效果

3.2.3 progress进度条

Progress进度条组件是一种提高用户体验度的组件,可以通过进度条看到完整视频的长度、当前播放的进度,这样让用户能合理地安排自己的时间,提高用户体验度。微信小程序也提供了progress进度条组件,它的属性如表3.6所示。

表3.6 进度条属性

属性	类型	默认值	说明
percent	Float	无	百分比为0~100
show-info	Boolean	false	在进度条右侧显示百分比
stroke-width	Number	6	进度条线的宽度,单位为px
Color	Color	#09BB07	进度条颜色
active	Boolean	false	进度条从左往右的动画

示例代码如下:

```
<progress percent="20" show-info />
<progress percent="40" stroke-width="12" />
<progress percent="60" color="pink" />
<progress percent="80" active />
```

界面效果如图3.12所示。

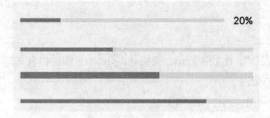

图3.12 进度条效果

3.2.4 rich-text富文本

rich-text富文本组件,可以在WXML页面文件显示一些富文本内容,比如显示HTML的一些元素内容。它有一个nodes节点列表属性,nodes 属性推荐使用 Array 类型,由于组件会将 String 类型转换为 Array 类型,因而性能会有所下降。nodes支持两种节点,通过type来区分,分别是元素节点和文本节点。

如表3.7和表3.8所示，它默认的是元素节点，即在富文本区域里显示的HTML节点。

1. 元素节点：type= node

表3.7 元素节点

属性	类型	默认值	说明
name	标签名	String	支持部分受信任的HTML节点
attrs	属性	Object	支持部分受信任的属性，遵循Pascal命名法
children	子节点列表	Array	结构和nodes一致

2. 文本节点：type= text

表3.8 文本节点

属性	类型	默认值	说明
text	文本	String	支持entities

示例代码如下：

```
<rich-text nodes="{{nodes}}" bindtap="tap"></rich-text>

Page({
  data: {
    nodes: [{
      name: 'div',
      attrs: {
        class: 'div_class',
        style: 'line-height: 60px; color: red;'
      },
      children: [{
        type: 'text',
        text: 'Hello World!'
      }]
    }]
  },
  tap() {
    console.log('tap')
  }
})
```

3.3 丰富的表单组件

微信小程序提供了丰富的表单组件：button按钮组件、checkbox多项选择器组件、radio单项选择器组件、input单行输入框组件、textarea多行输入框组件、label改进表单可用性组件、picker滚动选择器组件、slider滑动选择器组件、switch开关选择器组件、form表单组件10种表单组件。

丰富的表单组件

3.3.1 button按钮

button按钮组件提供3种类型按钮：基本类型按钮、默认类型按钮和警告类型按钮。同时提供两种大小形状的按钮：默认和mini两种大小按钮，如图3.13所示。

button按钮组件有很多属性，每个属性有不同的作用，如表3.9所示。

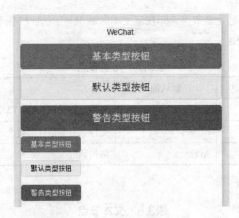

图3.13 按钮类型和大小

表3.9 按钮属性

属性	类型	默认值	说明
size	String	Default	有效值为：default、mini
type	String	Default	按钮的样式类型，有效值为primary、default、warn
plain	Boolean	False	按钮是否镂空，背景色透明
disabled	Boolean	False	是否禁用
loading	Boolean	False	名称前是否带 loading 图标
form-type	String	无	有效值为：submit、reset。用于 <form/> 组件，单击分别会触发 submit、reset 事件
hover-class	String	button-hover	指定按钮按下去的样式类。当hover-class="none" 时，没有单击态效果
hover-start-time	Number	50	按住后多久出现单击态，单位为毫秒
hover-stay-time	Number	400	手指松开后单击态保留时间，单位为毫秒

从按钮属性中可以看出按钮可以设置不同大小、不同类型、是否镂空、是否禁用、按钮名称前是否带loading图标。针对form表单组件，按钮组件提供了提交表单和重置表单两个功能，具体代码如下所示。

```
<button type="default" size="{{defaultSize}}" loading="{{loading}}" plain="{{plain}}"
    disabled="{{disabled}}" bindtap="default" > default </button>
<button type="primary" size="{{primarySize}}" loading="{{loading}}" plain="{{plain}}"
    disabled="{{disabled}}" bindtap="primary"> primary </button>
<button type="warn" size="{{warnSize}}" loading="{{loading}}" plain="{{plain}}"
    disabled="{{disabled}}" bindtap="warn"> warn </button>
<button bindtap="setDisabled">点击设置以上按钮disabled属性</button>
<button bindtap="setPlain">点击设置以上按钮plain属性</button>
<button bindtap="setLoading">点击设置以上按钮loading属性</button>

var types = ['default', 'primary', 'warn']
var pageObject = {
  data: {
    defaultSize: 'default',
    primarySize: 'default',
    warnSize: 'default',
    disabled: false,
```

```
      plain: false,
      loading: false
    },
    setDisabled: function(e) {
      this.setData({
        disabled: !this.data.disabled
      })
    },
    setPlain: function(e) {
      this.setData({
        plain: !this.data.plain
      })
    },
    setLoading: function(e) {
      this.setData({
        loading: !this.data.loading
      })
    }
}

for (var i = 0; i < types.length; ++i) {
  (function(type) {
    pageObject[type] = function(e) {
      var key = type + 'Size'
      var changedData = {}
      changedData[key] =
        this.data[key] === 'default' ? 'mini' : 'default'
      this.setData(changedData)
    }
  })(types[i])
}

Page(pageObject)
```

界面效果如图3.14所示。

图3.14　按钮效果

3.3.2　checkbox多项选择器

checkbox多项选择器组件，也就是我们常说的复选框，它用来进行多项选择，它的属性如表3.10所示。

表3.10　多项选择器属性

属性	类型	默认值	说明
value	String		\<checkbox/\>标识，选中时触发\<checkbox-group/\>的change事件，并携带 \<checkbox/\> 的 value
disabled	Boolean	False	是否禁用
checked	Boolean	False	当前是否选中，可用来设置默认选中
color	Color		checkbox的颜色，同css的color

checkbox-group是用来容纳多个checkbox多项选择器的容器，它有一个绑定事件bindchange，\<checkbox-group/\>中选中项发生改变时触发 change 事件，detail = {value:[选中的checkbox的value的数组]}。

下面演示一下checkbox多项选择器的使用，以及如何获取选中的value值。

（1）在wxml文件里使用checkbox进行界面布局，具体代码如下所示。

```
<checkbox-group bindchange="checkboxChange">
    <checkbox value="USA"/>美国
    <checkbox value="CHN" checked="true"/>中国
    <checkbox value="BRA"/>巴西
    <checkbox value="JPN"/>日本
    <checkbox value="ENG" disabled/>英国
</checkbox-group>
```

（2）在js文件里，添加checkboxChange 事件函数，获取复选框选中的值，将其打印出来，具体代码如下所示。

```
Page({
  checkboxChange:function(e){
      console.log(e.detail.value)
  }
})
```

界面效果如图3.15所示。

图3.15　复选框value值

从图3.15中可以看出，被禁用的复选框是不能使用的，在checkbox-group上面绑定bindchange事件，每次勾选时，会把所有勾选的复选框的值以数组的形式存在detail里。

3.3.3　radio单项选择器

radio单项选择器是与checkbox多项选择器对立的一个组件，每次只能选中一个radio单项选择器，选项间是一种互斥关系，它的属性如表3.11所示。

表3.11 单项选择器属性

属性	类型	默认值	说明
value	String		<radio/> 标识。当该<radio/> 选中时，<radio-group/> 的change 事件会携带<radio/>的value
disabled	Boolean	False	是否禁用
checked	Boolean	False	当前是否选中，可用来设置默认选中
color	Color		radio的颜色，同css的color

radio-group是用来容纳多个radio单项选择器的容器，它有一个绑定事件bindchange，<radio-group/> 中的选中项发生变化时触发 bindchange事件，event.detail = {value: 选中项radio的value}。

下面演示一下radio单项选择器的使用。

（1）在wxml文件里使用radio单项选择器进行界面布局，具体代码如下所示。

```
<radio-group class="radio-group" bindchange="radioChange">
    <radio value="USA" />美国
    <radio value="CHN" checked/>中国
    <radio value="BRA" disabled/>巴西
    <radio value="JPN" />日本
    <radio value="ENG" />英国
</radio-group>
```

（2）在js文件里，添加radioChange事件函数，获取单项选中的值，将其打印出来，具体代码如下所示。

```
Page({
    radioChange: function(e) {
        console.log('radio发生change事件，携带value值为：', e.detail.value)
    }
})
```

界面效果如图3.16所示。

图3.16 单项选择器value值

从图3.16中可以看出，被禁用的单项选择器是不能使用的，在radio-group上面绑定bindchange事件，每次勾选时，只能使一个选项呈现为选中状态，同时会把相应的值存在detail里。

3.3.4 input单行输入框

input单行输入框，用来输入单行文本内容，它的属性如表3.12所示。

表3.12 单行输入框属性

属性	类型	默认值	说明
Value	String		输入框的初始内容
Type	String	Text	input 的类型，有效值为text、number、idcard、digit

续表

属性	类型	默认值	说明
Password	Boolean	False	是否是密码类型
Placeholder	String		输入框为空时占位符
placeholder-style	String		指定 placeholder 的样式
placeholder-class	String	input-placeholder	指定 placeholder 的样式类
Disabled	Boolean	False	是否禁用
Maxlength	Number	140	最大输入长度，设置为 -1 的时候不限制最大长度
cursor-spacing	Number	0	指定光标与键盘的距离，单位为px。取input距离底部的距离和 cursor-spacing 指定的距离的最小值作为光标与键盘的距离
auto-focus	Boolean	False	（即将废弃，请直接使用focus）自动聚焦，拉起键盘
Focus	Boolean	False	获取焦点
Bindinput	EventHandle		当键盘输入时，触发input事件，event.detail = {value: value}，处理函数可以直接返回一个字符串，将替换输入框的内容
Bindfocus	EventHandle		输入框聚焦时触发，event.detail = {value: value}
Bindblur	EventHandle		输入框失去焦点时触发，event.detail = {value: value}
Bindconfirm	EventHandle		单击完成按钮时触发，event.detail = {value: value}

从表3.12单行输入框属性里可以看出：

（1）可以设置input输入框的类型有text、number、idcard、digit；

（2）可以设置输入框是否为密码类型，如果是密码类型，则会用点号代替具体值显示；

（3）通过placeholder来给输入框添加类似于"请输入手机号/用户名/邮箱"这样友好的提示信息，placeholder-style设置提示信息的样式，placeholder-class设置提示信息的class，然后再针对这个class添加样式；

（4）可以设置input输入框禁用、最大长度和获取焦点；

（5）input输入框有3个常用的事件：输入时（bindinput）、光标聚焦时（bindfocus）、光标离开时（bindblur）。

示例代码如下：

（1）在wxml文件里利用input单行输入框进行布局，具体代码如下所示。

```
<view class="section">
  <input placeholder="这是一个可以自动聚焦的input" auto-focus/>
</view>
<view class="section">
  <input placeholder="这个只有在按钮点击的时候才聚焦" focus="{{focus}}" />
  <view class="btn-area">
    <button bindtap="bindButtonTap">使得输入框获取焦点</button>
  </view>
</view>
<view class="section">
  <input maxlength="10" placeholder="最大输入长度10" />
</view>
```

```
<view class="section">
  <view class="section__title">你输入的是：{{inputValue}}</view>
  <input  bindinput="bindKeyInput" placeholder="输入同步到view中"/>
</view>
<view class="section">
  <input  bindinput="bindReplaceInput" placeholder="连续的两个1会变成2" />
</view>
<view class="section">
  <input  bindinput="bindHideKeyboard" placeholder="输入123自动收起键盘" />
</view>
<view class="section">
  <input password type="number" />
</view>
<view class="section">
  <input password type="text" />
</view>
<view class="section">
  <input type="digit" placeholder="带小数点的数字键盘"/>
</view>
<view class="section">
  <input type="idcard" placeholder="身份证输入键盘" />
</view>
<view class="section">
  <input placeholder-style="color:red" placeholder="占位符字体是红色的" />
</view>
```

（2）在js文件里给input单行输入框添加相应的事件并提供数据，具体代码如下所示。

```
Page({
  data: {
    focus: false,
    inputValue: ''
  },
  bindButtonTap: function() {
    this.setData({
      focus: true
    })
  },
  bindKeyInput: function(e) {
    this.setData({
      inputValue: e.detail.value
    })
  },
  bindReplaceInput: function(e) {
    var value = e.detail.value
    var pos = e.detail.cursor
    if(pos != -1){
      //光标在中间
      var left = e.detail.value.slice(0,pos)
      //计算光标的位置
      pos = left.replace(/11/g,'2').length
    }

    //直接返回对象，可以对输入进行过滤处理，同时可以控制光标的位置
```

```
      return {
        value: value.replace(/11/g,'2'),
        cursor: pos
      }

      //或者直接返回字符串,光标在最后边
      //return value.replace(/11/g,'2'),
    },
    bindHideKeyboard: function(e) {
      if (e.detail.value === '123') {
        //收起键盘
        wx.hideKeyboard()
      }
    }
})
```

界面效果如图3.17所示。

图3.17　input单行输入框

input 组件是一个 native 组件，字体是系统字体，所以无法设置 font-family；在 input 聚焦期间，避免使用 css 动画。

3.3.5　textarea多行输入框

textarea多行输入框是与input单行输入框对应的一个组件，它可以输入多行文本内容，它的属性如表3.13所示。

表3.13　多行输入框属性

属性	类型	默认值	说明
value	String		输入框的内容
placeholder	String		输入框为空时占位符
placeholder-style	String		指定 placeholder 的样式
placeholder-class	String	input-placeholder	指定 placeholder 的样式类

续表

属性	类型	默认值	说明
disabled	Boolean	False	是否禁用
maxlength	Number	140	最大输入长度，设置为 -1 的时候不限制最大长度
cursor-spacing	Number	0	指定光标与键盘的距离，单位为px。取textarea距离底部的距离和cursor-spacing指定的距离的最小值作为光标与键盘的距离
auto-focus	Boolean	False	（即将废弃，请直接使用 focus）自动聚焦，拉起键盘
focus	Boolean	False	获取焦点
auto-height	Boolean	False	是否自动增高，设置auto-height时，style.height不生效
fixed	Boolean	False	如果 textarea 是在一个 position:fixed 的区域，需要显示指定属性 fixed 为 true
bindlinechange	EventHandle		输入框行数变化时调用，event.detail = {height: 0, heightRpx: 0, lineCount: 0}
bindinput	EventHandle		当键盘输入时，触发input事件，event.detail = {value: value}，处理函数可以直接返回一个字符串，将替换输入框的内容
bindfocus	EventHandle		输入框聚焦时触发，event.detail = {value: value}
bindblur	EventHandle		输入框失去焦点时触发，event.detail = {value: value}
bindconfirm	EventHandle		单击完成按钮时触发，event.detail = {value: value}

从表3.13多行输入框属性里可以看出：

（1）通过placeholder来给输入框添加类似于"请输入手机号/用户名/邮箱"这样友好的提示信息，placeholder-style设置提示信息的样式，placeholder-class设置提示信息的类，然后再针对这个类添加样式；

（2）可以设置textarea输入框禁用和最大长度、获取焦点、自动调整行高；

（3）input输入框有4个常用的事件：输入时（bindinput）、光标聚焦时（bindfocus）、光标离开时（bindblur）、行数变化时（bindlinechange）。

示例代码如下所示：

```
<view class="section">
  <textarea bindblur="bindTextAreaBlur" auto-height placeholder="自动变高" />
</view>
<view class="section">
  <textarea placeholder="placeholder颜色是红色的" placeholder-style="color:red;" />
</view>
```

请勿在 scroll-view 中使用 textarea 组件，textarea 的 blur 事件会晚于页面上的 tap 事件，如果需要在 button 的单击事件中获取 textarea，可以使用 form 的bindsubmit，不建议在多行文本上对用户的输入进行修改，所以 textarea 的 bindinput 处理函数并不会将返回值反映到 textarea 上。

3.3.6 label改进表单可用性

label组件是用来改进表单可用性的，目前可以用来改进的组件有：<button/>、<checkbox/>、

<radio/>、<switch/>。它只有一个属性for，属性for用来绑定控件的id。它有两种使用方式：一种是没有定义for属性，另一种是定义for属性。

1. label组件没有定义for属性

label组件没有定义for属性时，在label内包含<button/>、<checkbox/>、<radio/>、<switch/>这些组件，当单击label组件时，会触发label内包含的第一个控件，假如<button/>在第一个位置，就会触发<button/>对应的事件，假如<radio/>在第一个位置，就会触发radio对应的事件。

下面演示一下它的使用。

（1）在wxml文件里利用label组件布局，把第一个组件隐藏起来，具体代码如下所示。

```
<label>
    <button bindtap="clickBtn" hidden>我是button按钮</button>
    <view>我是label组件内的内容</view>
    <checkbox-group bindchange="checkboxChange">
        <checkbox value="中国" />中国
        <checkbox value="美国" />美国
    </checkbox-group>
    <radio-group bindchange="radioChange">
        <radio value="男"/>男
        <radio value="女"/>女
    </radio-group>
</label>
```

（2）在js文件里添加clickBtn、checkboxChange、radioChange 3个事件函数，分别打印不同的信息，具体代码如下所示。

```
Page({
    clickBtn:function(){
        console.log("单击了按钮组件");
    },
    checkboxChange:function(){
        console.log("单击了多项选择器组件");
    },
    radioChange:function(){
        console.log("单击了单选选择器组件");
    }
})
```

（3）在wxml界面里可以看到<button/>按钮组件是隐藏起来的，但是单击"我是label组件内的内容"，可以看到打印信息是按钮事件函数打印的信息，如图3.18所示。

图3.18 没有定义for属性

从这里可以看出，label组件内有多个组件时，会触发第一个组件。

2. label组件定义for属性

label组件定义for属性后，它会根据for属性的值找到与组件id一样的值，然后触发这个组件的相应事件。

下面演示一下它的使用：

（1）在wxml文件里利用label组件布局，把第一个组件隐藏起来，给label定义for等于man，让它找到id值等于man组件，然后触发该组件的事件，具体代码如下所示。

```
<label for="man">
    <button id="btn" bindtap="clickBtn" hidden>我是button按钮</button>
    <view>我是label组件内的内容</view>
    <checkbox-group bindchange="checkboxChange" id="checkbox">
        <checkbox value="中国" />中国
        <checkbox value="美国" />美国
    </checkbox-group>
    <radio-group bindchange="radioChange" >
        <radio id="man" value="男"/>男
        <radio id="women" value="女"/>女
    </radio-group>
</label>
```

（2）在js文件里添加clickBtn、checkboxChange、radioChange 3个事件函数，分别打印不同的信息，具体代码如下所示。

```
Page({
    clickBtn:function(){
        console.log("单击了按钮组件");
    },
    checkboxChange:function(){
        console.log("单击了多项选择器组件");
    },
    radioChange:function(){
        console.log("单击了单选选择器组件");
    }
})
```

（3）在wxml界面里可以看到<button/>按钮组件是隐藏起来的，但是单击"我是label组件内的内容"，可以看到id值等于man的单项选择器程序为选中状态，同时触发事件，打印信息，如图3.19所示。

图3.19　定义for属性

从这里可以看出，如果label定义for属性，就会根据for属性的值找到组件id等于for属性值，然后触发相应事件；如果label没有定义for属性，它会找到label组件内的第一个组件，然后触发相应事件。

3.3.7　picker滚动选择器

picker滚动选择器，支持3种滚动选择器：普通选择器、时间选择器、日期选择器。默认的是普通选择器，如图3.20、图3.21和图3.22所示。

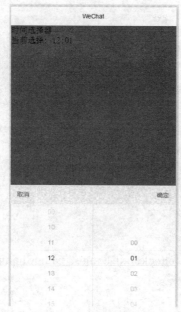
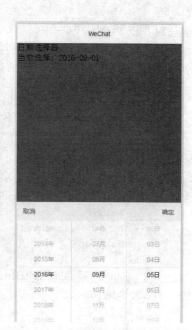

图3.20　普通选择器　　　　　图3.21　时间选择器　　　　　图3.22　日期选择器

这3种选择器是通过mode来区分的，普通选择器mode = selector，时间选择器mode = time，日期选择器mode = date，每种类型选择器的属性不同，如表3.14、表3.15、表3.16所示。

1. **普通选择器：mode = selector**

表3.14　普通选择器属性

属性	类型	默认值	说明
Range	Array / Object Array	[]	mode为 selector 时，range 有效
range-key	String		当 range 是一个 Object Array 时，通过 range-key 来指定 Object 中 key 的值作为选择器显示内容
Value	Numbe	0	value 的值表示选择了 range 中的第几个（下标从 0 开始）
Bindchange	EventHandle		value 改变时触发 change 事件，event.detail = {value: value}
disabled	Boolean	false	是否禁用

示例代码如下：

```
<view class="section">
  <view class="section__title">地区选择器</view>
  <picker bindchange="bindPickerChange" value="{{index}}" range="{{array}}">
    <view class="picker">
      当前选择：{{array[index]}}
    </view>
  </picker>
</view>


Page({
  data: {
    array: ['美国', '中国', '巴西', '日本'],
    objectArray: [
```

```
      {
        id: 0,
        name: '美国'
      },
      {
        id: 1,
        name: '中国'
      },
      {
        id: 2,
        name: '巴西'
      },
      {
        id: 3,
        name: '日本'
      }
    ],
    index: 0
  },
  bindPickerChange: function(e) {
    console.log('picker发送选择改变，携带值为', e.detail.value)
    this.setData({
      index: e.detail.value
    })
  }
})
```

界面效果如图3.23所示。

图3.23　普通选择器

2. 时间选择器：mode = time

表3.15 时间选择器属性

属性	类型	默认值	说明
value	String		表示选中的时间，格式为"hh:mm"
start	String		表示有效时间范围的开始，字符串格式为"hh:mm"
end	String		表示有效时间范围的结束，字符串格式为"hh:mm"
bindchange	EventHandle		value 改变时触发 change 事件，event.detail = {value: value}
disabled	Boolean	False	是否禁用

示例代码如下：

```
<view class="section">
  <view class="section__title">时间选择器</view>
  <picker mode="time" value="{{time}}" start="09:01" end="21:01" bindchange="bindTimeChange">
    <view class="picker">
      当前选择: {{time}}
    </view>
  </picker>
</view>

Page({
  data: {
    time: '12:01'
  },
  bindTimeChange: function(e) {
    this.setData({
      time: e.detail.value
    })
  }
})
```

界面效果如图3.24所示。

图3.24 时间选择器

3. 日期选择器：mode = date

表3.16 日期选择器属性

属性	类型	默认值	说明
value	String	0	表示选中的日期，格式为"YYYY-MM-DD"
start	String		表示有效日期范围的开始，字符串格式为"YYYY-MM-DD"
end	String		表示有效日期范围的结束，字符串格式为"YYYY-MM-DD"
fields	String	Day	有效值为year、month、day，表示选择器的粒度
bindchange	EventHandle		value 改变时触发 change 事件，event.detail = {value: value}
disabled	Boolean	False	是否禁用

示例代码如下：

```
<view class="section">
  <view class="section__title">日期选择器</view>
  <picker mode="date" value="{{date}}" start="2015-09-01" end="2017-09-01" bindchange="bindDateChange">
    <view class="picker">
      当前选择: {{date}}
    </view>
  </picker>
</view>

Page({
  data: {
    date: '2016-09-01'
  },
  bindDateChange: function(e) {
    this.setData({
      date: e.detail.value
    })
  }
})
```

界面效果如图3.25所示。

图3.25 日期选择器

4. picker-view嵌入页面滚动选择器

除了普通选择器、时间选择器、日期选择器3种滚动选择器之外,还有一种嵌入页面的滚动选择器,它使用picker-view组件在页面里的布局,如图3.26所示。

图3.26 嵌入页面滚动选择器

picker-view嵌入页面滚动选择器组件里面只能显示<picker-view-column/>组件,不会显示其他节点,picker-view有3个属性,如表3.17所示。

表3.17 嵌入页面滚动选择器属性

属性	类型	默认值	说明
Value	Number	Array	数组中的数字依次表示 picker-view 内的 picker-view-colume 选择的第几项(下标从 0 开始),数字大于 picker-view-column 可选项长度时,选择最后一项
indicator-style	String		设置选择器中间选中框的样式
bindchange	EventHandle		当滚动选择,value 改变时触发 change 事件,event.detail = {value: value}; value为数组,表示 picker-view 内的 picker-view-column 当前选择的是第几项(下标从 0 开始)

示例代码如下所示:

```
<view>
  <view style="text-align:center">{{year}}年{{month}}月{{day}}日</view>
  <picker-view indicator-style="height: 50px;" style="width: 100%; height: 300px;" value="{{value}}" bindchange="bindChange">
    <picker-view-column>
      <view wx:for="{{years}}" style="line-height: 50px">{{item}}年</view>
    </picker-view-column>
    <picker-view-column>
      <view wx:for="{{months}}" style="line-height: 50px">{{item}}月</view>
    </picker-view-column>
    <picker-view-column>
      <view wx:for="{{days}}" style="line-height: 50px">{{item}}日</view>
    </picker-view-column>
  </picker-view>
</view>

const date = new Date()
const years = []
```

```
const months = []
const days = []

for (let i = 1990; i <= date.getFullYear(); i++) {
  years.push(i)
}

for (let i = 1 ; i <= 12; i++) {
  months.push(i)
}

for (let i = 1 ; i <= 31; i++) {
  days.push(i)
}

Page({
  data: {
    years: years,
    year: date.getFullYear(),
    months: months,
    month: 2,
    days: days,
    day: 2,
    year: date.getFullYear(),
    value: [9999, 1, 1],
  },
  bindChange: function(e) {
    const val = e.detail.value
    this.setData({
      year: this.data.years[val[0]],
      month: this.data.months[val[1]],
      day: this.data.days[val[2]]
    })
  }
})
```

3.3.8 slider滑动选择器

slider滑动选择器组件，经常用于控制声音的大小、屏幕的亮度等场景，它可以设置滑动步长、显示当前值以及设置最小值、最大值，如图3.27所示。

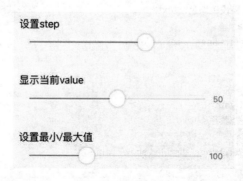

图3.27 滑动选择器

slider滑动选择器组件的属性如表3.18所示。

表3.18 slider滑动选择器属性

属性	类型	默认值	说明
Min	Number	0	最小值
Max	Number	100	最大值
Step	Number	1	步长，取值必须大于 0，并且可被(max – min)整除
disabled	Boolean	False	是否禁用
value	Number	0	当前取值
color	Color	#e9e9e9	背景条的颜色
selected-color	Color	#1aad19	已选择的颜色
show-value	Boolean	False	是否显示当前 value
bindchange	EventHandle		完成一次拖动后触发的事件，event.detail = {value: value}

示例代码如下所示：

```
<view class="section section_gap">
  <text class="section__title">设置step</text>
  <view class="body-view">
    <slider bindchange="sliderchange" step="5"/>
  </view>
</view>

<view class="section section_gap">
  <text class="section__title">显示当前value</text>
  <view class="body-view">
    <slider bindchange="sliderchange" show-value/>
  </view>
</view>

<view class="section section_gap">
  <text class="section__title">设置最小/最大值</text>
  <view class="body-view">
    <slider bindchange="sliderchange" min="50" max="200" show-value/>
  </view>
</view>

<view class="section section_gap">
  <text class="section__title">设置颜色</text>
  <view class="body-view">
    <slider bindchange="sliderchange"  color="black" selected-color="red"/>
  </view>
</view>

<view class="section section_gap">
  <text class="section__title">禁用</text>
  <view class="body-view">
    <slider bindchange="sliderchange" disabled show-value/>
  </view>
</view>
```

界面效果如图3.28所示。

图3.28　滑动选择器使用

3.3.9　switch开关选择器

switch开关选择器应用得十分普遍，它有两个状态：开和关。在很多场景都会用到开关这个功能，比如微信设置里的新消息提醒界面，它通过开关来设置是否接收消息、是否显示消息、是否有声音、是否振动等功能，如图3.29所示。

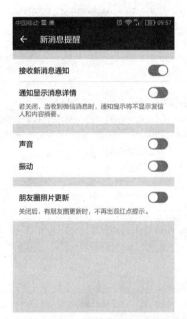

图3.29　微信新消息提醒设置

switch开关选择器的属性可以设置为是否选中、开关类型、绑定事件和颜色，如表3.19所示。

表3.19　switch开关选择器属性

属性	类型	默认值	说明
checked	Boolean	False	是否选中
type	String	Switch	样式，有效值为：switch, checkbox
bindchange	EventHandle		checked 改变时触发 change 事件，event.detail={ value:checked }
color	Color		switch 的颜色，同 css 的 color

示例代码如下所示:

```html
<view style="background-color:#cccccc;height:600px;">
  <view style="padding-top:10px;"></view>
  <view style="display:flex;flex-direction:row;background-color:#ffffff;height:50px;line-height:50px;">
    <view style="font-weight:bold;">接收新消息通知</view>
    <view style="position:absolute;right:10px;">
      <switch type="switch" checked/>
    </view>
  </view>
  <view style="height:1px;background-color:#f2f2f2;opacity:0.2"></view>
  <view style="display:flex;flex-direction:row;background-color:#ffffff;height:50px;line-height:50px;">
    <view style="font-weight:bold;">通知显示消息详情</view>
    <view style="position:absolute;right:10px;">
      <switch type="switch"/>
    </view>
  </view>
  <view style="height:1px;background-color:#f2f2f2;opacity:0.2"></view>

  <view style="margin-top:20px;"></view>
  <view style="height:1px;background-color:#f2f2f2;opacity:0.2"></view>
  <view style="display:flex;flex-direction:row;background-color:#ffffff;height:50px;line-height:50px;">
    <view style="font-weight:bold;">声音</view>
    <view style="position:absolute;right:10px;">
      <switch type="checkbox" checked/>
    </view>
  </view>
  <view style="height:1px;background-color:#f2f2f2;opacity:0.2"></view>
  <view style="height:1px;background-color:#f2f2f2;opacity:0.2"></view>
  <view style="display:flex;flex-direction:row;background-color:#ffffff;height:50px;line-height:50px;">
    <view style="font-weight:bold;">振动</view>
    <view style="position:absolute;right:10px;">
      <switch type="checkbox"/>
    </view>
  </view>
  <view style="height:1px;background-color:#f2f2f2;opacity:0.2"></view>
</view>
```

界面效果如图3.30所示。

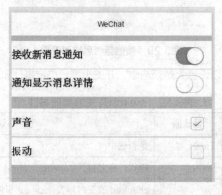

图3.30 开关选择器应用

3.3.10 form表单

form表单组件将表单里组件的值提交给js文件进行处理,可以提交<switch/>、<input/>、<checkbox/>、<slider/>、<radio/>、<picker/>这些组件的值。提交表单的时候,会借助于button组件的formType为submit的属性,将表单组件中的 value 值进行提交,需要在表单组件中加上 name 来作为key。form表单组件的属性如表3.20所示。

表3.20 form表单属性

属性	类型	默认值	说明
report-submit	Boolean		是否返回 formId 用于发送模板消息
bindsubmit	EventHandle		携带 form 中的数据触发 submit 事件,event.detail = {value : {'name': 'value'} , formId: ''}
bindreset	EventHandle		表单重置时会触发 reset 事件

示例代码如下所示:

```
<form bindsubmit="formSubmit" bindreset="formReset">
  <view style="margin:10px;">
    <view style="font-weight:bold;">switch开关选择器</view>
    <switch name="switch"/>
  </view>
  <view style="margin:10px;">
    <view style="font-weight:bold;">slider滑动选择器</view>
    <slider name="slider" show-value ></slider>
  </view>

  <view style="margin:10px;">
    <view style="font-weight:bold;">input单行输入框</view>
    <input name="input" placeholder="please input here" />
  </view>
  <view style="margin:10px;">
    <view style="font-weight:bold;">radio单项选择器</view>
    <radio-group name="radio-group">
      <label><radio value="radio1"/>radio1</label>
      <label><radio value="radio2"/>radio2</label>
    </radio-group>
  </view>
  <view style="margin:10px;">
    <view style="font-weight:bold;">checkbox多项选择器</view>
    <checkbox-group name="checkbox">
      <label><checkbox value="checkbox1"/>checkbox1</label>
      <label><checkbox value="checkbox2"/>checkbox2</label>
    </checkbox-group>
  </view>
  <view class="btn-area">
    <button formType="submit" type="primary">Submit</button>
    <button formType="reset">Reset</button>
  </view>
</form>

Page({
```

```
formSubmit: function(e) {
    console.log('form发生了submit事件，携带数据为：', e.detail.value)
},
formReset: function() {
    console.log('form发生了reset事件')
}
})
```

界面效果如图3.31、图3.32所示。

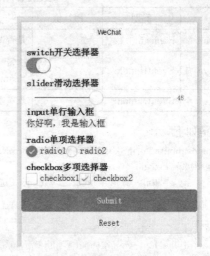

图3.31　未填写表单　　　　　　图3.32　填写表单

单击Reset按钮可以重置表单，单击Submit按钮组件，就可以把表单数据提交到js文件里进行处理，提交数据如图3.33所示。

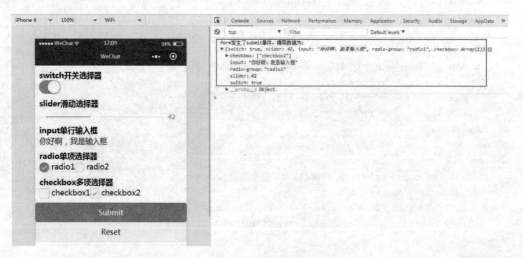

图3.33　表单提交数据

3.4　导航组件

微信小程序可以在页面中设置导航，可以使用navigator页面链接组件，也可以在js文件里设置导航进行页面跳转，同时可以设置导航条标题和显示动画效果。

导航组件

3.4.1 navigator页面链接组件

navigator页面链接组件用来在WXML页面中实现跳转，它有3种类型：第1种是保留当前页跳转，跳转后可以返回当前页，它与wx.navigateTo跳转效果是一样的；第2种是关闭当前页跳转，无法返回当前页，它与wx.redirectTo跳转效果是一样的；第3种是跳转到底部标签导航指定的页面，它与wx.switchTab跳转效果是一样的。navigator页面链接组件的这些跳转效果都是通过open-type属性来控制的，具体属性如表3.21所示。

表3.21 navigato页面链接属性

属性	类型	默认值	说明
url	String		应用内的跳转链接
redirect	Boolean	False	打开方式为页面重定向，对应 wx.redirectTo（将被废弃，推荐使用open-type）
open-type	String	Navigate	可选值'navigate'、'redirect'、'switchTab'，对应于wx.navigateTo、wx.redirectTo、wx.switchTab的功能
hover-class	String	navigator-hover	指定单击时的样式类，当hover-class="none"时，没有单击态效果
hover-start-time	Number	50	按住后多久出现单击态，单位为毫秒
hover-stay-time	Number	600	手指松开后单击态保留时间，单位为毫秒

下面来演示一下open-type不同导航类型的跳转效果。

（1）新建一个navigator项目，进入到app.json文件中，在pages属性里设置页面路径"pages/index/index""pages/navigator/navigator""pages/redirect/redirect"，具体代码如下所示。

```
{
  "pages":[
    "pages/index/index",
    "pages/navigator/navigator",
    "pages/redirect/redirect"
  ],
  "window":{
    "backgroundTextStyle":"light",
    "navigationBarBackgroundColor": "#fff",
    "navigationBarTitleText": "导航",
    "navigationBarTextStyle":"black"
  },
}
```

（2）进入到pages/index/index.wxml文件里，设置导航的3种跳转方式：保留当前页跳转、关闭当前页跳转、跳转到tabBar页面，具体代码如下所示。

```
<view class="btn-area">
    <navigator url="../navigator/navigator?title=navigator" open-type="navigate" hover-class="navigator-hover">wx.navigateTo保留当前页跳转</navigator>
    <navigator url="../redirect/redirect?title=redirect" open-type="redirect" hover-class="other-navigator-hover">wx.redirectTo关闭当前页跳转</navigator>
    <navigator url="../redirect/redirect" open-type="switchTab" hover-class="other-navigator-hover">wx.switchTab跳转到tabBar页面</navigator>
</view>
```

（3）进入到pages/navigator/navigator.wxml文件里，进行界面布局，具体代码如下所示。

`<view>`保留当前页进行跳转，单击左上角可以返回到当前页`</view>`

（4）进入到pages/redirect/redirect.wxml文件里，进行界面布局，具体代码如下所示。

`<view>`关闭当前页进行跳转，跳转后无法返回到当前页 `</view>`

（5）wx.navigateTo保留当前页跳转、wx.redirectTo关闭当前页跳转这两种方式可以正常跳转，但是wx.switchTab跳转到tabBar页面这种方式是无法正常跳转的，它需要在app.json文件的tabBar属性里设置底部标签导航，具体代码如下所示。

```
{
  "pages":[
    "pages/index/index",
    "pages/navigator/navigator",
    "pages/redirect/redirect"
  ],
  "window":{
    "backgroundTextStyle":"light",
    "navigationBarBackgroundColor": "#fff",
    "navigationBarTitleText": "导航",
    "navigationBarTextStyle":"black"
  },
  "tabBar": {
    "selectedColor": "red",
    "list": [{
      "pagePath": "pages/index/index",
      "text": "首页",
      "iconPath": "iconPath",
      "selectedIconPath": "selectedIconPath"
    },{
      "pagePath": "pages/redirect/redirect",
      "text": "当前页打开导航",
      "iconPath": "iconPath",
      "selectedIconPath": "selectedIconPath"
    }]
  }
}
```

（6）使用wx.switchTab跳转到tabBar页面这种方式可以正常跳转到指定的底部标签导航页面里，但是wx.navigateTo保留当前页跳转、wx.redirectTo关闭当前页跳转这两种方式无法跳转，这是因为在app.json中配置的tabBar属性里设置了底部标签导航。

（7）navigator页面链接组件设置的跳转路径，如果带参数，像url="../navigator/navigator?title=navigator"，获取title的值，可以在跳转页面里js文件的onLoad函数里获得，具体代码如下所示。

```
Page({
  data:{},
  onLoad:function(options){
    console.log("title="+options);
  }
})
```

3.4.2　wx.navigateTo保留当前页跳转

wx.navigateTo保留当前页面跳转到应用内的某个页面，使用wx.navigateBack可以返回到原页面，具体属性如表3.22所示。

表3.22 wx.navigateTo属性

属性	类型	是否必填	说明
url	String	是	需要跳转的应用内非tabBar的页面的路径，路径后可以带参数。参数与路径之间使用?分隔，参数键与参数值用=相连，不同参数用&分隔；如 'path?key=value&key2=value2'
success	Function	否	接口调用成功的回调函数
fail	Function	否	接口调用失败的回调函数
complete	Function	否	接口调用结束的回调函数（调用成功、失败都会执行）

（1）进入到pages/index/index.wxml文件里，添加一个跳转按钮，保留当前页进行跳转，具体代码如下所示。

```
<view class="btn-area">
    <navigator url="../navigator/navigator?title=navigator11" open-type="navigate" hover-class="navigator-hover">wx.navigateTo保留当前页跳转</navigator>
    <navigator url="../redirect/redirect?title=redirect" open-type="redirect" hover-class="other-navigator-hover">wx.redirectTo关闭当前页跳转</navigator>
    <navigator url="../redirect/redirect" open-type="switchTab" hover-class="other-navigator-hover">wx.switchTab跳转到tabBar页面</navigator>
    <button type="primary" bindtap="navigateBtn">保留当前页跳转</button>
</view>
```

（2）进入到pages/index/index.js文件里，添加一个navigateBtn事件函数，保留当前页跳转到pages/navigator/navigator.wxml页面里，具体代码如下所示。

```
Page({
    navigateBtn:function(){
    wx.navigateTo({
        url: '../navigator/navigator',
        success: function(res){
          console.log(res);
        },
        fail: function() {
          // fail
        },
        complete: function() {
          // complete
        }
    })
  }
})
```

3.4.3 wx.redirectTo关闭当前页跳转

wx.redirectTo关闭当前页面，跳转到应用内的某个页面，具体属性如表3.23所示。

表3.23 wx.redirectTo属性

属性	类型	是否必填	说明
url	String	是	需要跳转的应用内非 tabBar 的页面的路径，路径后可以带参数。参数与路径之间使用?分隔，参数键与参数值用=相连，不同参数用&分隔；如 'path?key=value&key2=value2'
success	Function	否	接口调用成功的回调函数

续表

属性	类型	是否必填	说明
fail	Function	否	接口调用失败的回调函数
complete	Function	否	接口调用结束的回调函数（调用成功、失败都会执行）

（1）进入到pages/index/index.wxml文件里，添加一个跳转按钮，关闭当前页进行跳转，具体代码如下所示。

```
<view class="btn-area">
  <navigator url="../navigator/navigator?title=navigator11" open-type="navigate" hover-class="navigator-hover">wx.navigateTo保留当前页跳转</navigator>
  <navigator url="../redirect/redirect?title=redirect" open-type="redirect" hover-class="other-navigator-hover">wx.redirectTo关闭当前页跳转</navigator>
  <navigator url="../redirect/redirect" open-type="switchTab" hover-class="other-navigator-hover">wx.switchTab跳转到tabBar页面</navigator>
  <button type="primary" bindtap="navigateBtn">保留当前页跳转</button>
  <button type="primary" bindtap="redirectBtn">关闭当前页跳转</button>
</view>
```

（2）进入到pages/index/index.js文件里，添加一个redirectBtn事件函数，保留当前页跳转到pages/navigator/navigator.wxml页面里，具体代码如下所示。

```
Page({
  navigateBtn:function(){
    wx.navigateTo({
      url: '../navigator/navigator',
      success: function(res){
        console.log(res);
      },
      fail: function() {
        // fail
      },
      complete: function() {
        // complete
      }
    })
  },
  redirectBtn:function(){
    wx.redirectTo({
      url: '../navigator/navigator',
      success: function(res){
        console.log(res);
      },
      fail: function() {
        // fail
      },
      complete: function() {
        // complete
      }
    })
  }
})
```

3.4.4 跳转到tabBar页面

wx.switchTab跳转到 tabBar 页面，并关闭其他所有非 tabBar 页面，具体属性如表3.24所示。

表3.24 wx.switchTab属性

属性	类型	是否必填	说明
url	String	是	需要跳转的 tabBar 页面的路径（需在 App.json 的 tabBar 字段定义的页面），路径后不能带参数
success	Function	否	接口调用成功的回调函数
fail	Function	否	接口调用失败的回调函数
complete	Function	否	接口调用结束的回调函数（调用成功、失败都会执行）

（1）进入到pages/index/index.wxml文件里，添加一个跳转按钮，跳转到tabBar页面，具体代码如下所示。

```
<view class="btn-area">
  <navigator url="../navigator/navigator?title=navigator11" open-type="navigate" hover-class="navigator-hover">wx.navigateTo保留当前页跳转</navigator>
  <navigator url="../redirect/redirect?title=redirect" open-type="redirect" hover-class="other-navigator-hover">wx.redirectTo关闭当前页跳转</navigator>
  <navigator url="../redirect/redirect" open-type="switchTab" hover-class="other-navigator-hover">wx.switchTab跳转到tabBar页面</navigator>
  <button type="primary" bindtap="navigateBtn">保留当前页跳转</button>
  <button type="primary" bindtap="redirectBtn">关闭当前页跳转</button>
  <button type="primary" bindtap="switchBtn">跳转到tabBar页面</button>
</view>
```

（2）进入到pages/index/index.js文件里，添加一个navigateBtn事件函数，保留当前页跳转到pages/redirect/redirect.wxml页面里，具体代码如下所示。

```
Page({
  navigateBtn:function(){
    wx.navigateTo({
      url: '../navigator/navigator',
      success: function(res){
        console.log(res);
      },
      fail: function() {
        // fail
      },
      complete: function() {
        // complete
      }
    })
  },
  redirectBtn:function(){
    wx.redirectTo({
      url: '../navigator/navigator',
      success: function(res){
        console.log(res);
      },
      fail: function() {
```

```
          // fail
        },
        complete: function() {
          // complete
        }
      })
    },
    switchBtn:function(){
      wx.switchTab({
        url: '../redirect/redirect',
        success: function(res){
          // success
        },
        fail: function() {
          // fail
        },
        complete: function() {
          // complete
        }
      })
    }
})
```

wx.navigateTo 和 wx.redirectTo 不允许跳转到 tabBar 页面，只能用 wx.switchTab 跳转到 tabBar 页面。

3.4.5 wx.navigateBack返回上一页

wx.navigateBack关闭当前页面，返回上一页面或多级页面。可通过 getCurrentPages()获取当前的页面栈，决定需要返回几层。具体属性如表3.25所示。

表3.25　wx.navigateBack属性

属性	类型	是否必填	说明
delta	Number	1	返回的页面数，如果 delta 大于现有页面数，则返回到首页

（1）进入到pages/navigator/navigator.wxml文件里，添加一个返回按钮，单击返回按钮，可以返回到上一级页面，具体代码如下所示。

```
<view>保留当前页进行跳转，单击左上角可以返回到当前页</view>
<button type="primary" bindtap="backBtn">返回上一页</button>
```

（2）进入到pages/navigator/navigator.js文件里，添加backBtn事件返回函数，具体代码如下所示。

```
Page({
  data:{},
  onLoad:function(options){
    console.log("title="+options);
  },
  backBtn:function(){
    wx.navigateBack({
      delta: 1
    })
  }
})
```

（3）在pages/index/index.wxml文件里，单击保留当前页跳转按钮，可以进行页面跳转，在跳转的页面里单击返回上一页按钮，可以返回到上一级页面，如图3.34、图3.35所示。

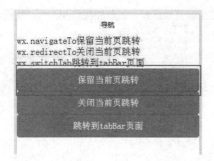

图3.34　index.wxml页面

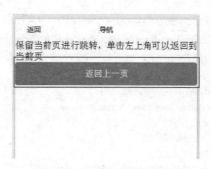

图3.35　navigator.wxml页面

3.4.6　设置导航条

wx.setNavigationBarTitle(OBJECT)动态设置当前页面的标题。具体属性如表3.26所示。

表3.26　wx.setNavigationBarTitle(OBJECT)属性

属性	类型	是否必填	说明
title	String	是	页面标题
success	Function	否	接口调用成功的回调函数
fail	Function	否	接口调用失败的回调函数
complete	Function	否	接口调用结束的回调函数（调用成功、失败都会执行）

（1）进入到pages/navigator/navigator.js文件里，对页面窗口标题重新设计，具体代码如下所示。

```
Page({
  data:{},
  onLoad:function(options){
    console.log("title="+options);
    wx.setNavigationBarTitle({
      title: '新页面'
    })
  },
  backBtn:function(){
    wx.navigateBack({
      delta: 1
    })
  }
})
```

（2）在pages/index/index.wxml文件里，单击保留当前页跳转按钮，可以进行页面跳转，跳转后标题变为新页面，如图3.36、图3.37所示。

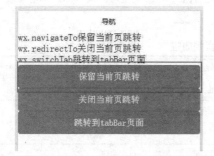

图3.36　index.wxml页面

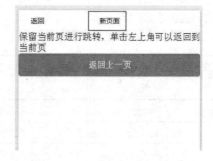

图3.37　navigator.wxml页面

（3）在pages/navigator/navigator.js文件里，如果在当前页面显示导航条加载动画，可以使用wx.showNavigationBarLoading()函数，具体代码如下所示。

```
Page({
  data:{},
  onLoad:function(options){
    console.log("title="+options);
    wx.setNavigationBarTitle({
      title: '新页面'
    });
    wx.showNavigationBarLoading();
  },
  backBtn:function(){
    wx.navigateBack({
      delta: 1
    })
  }
})
```

界面效果如图3.38所示。

图3.38　导航加载效果

（4）在当前页面显示导航条加载动画，可以使用wx.showNavigationBarLoading()函数，如果想隐藏导航条加载动画，可以使用wx.hideNavigationBarLoading()函数。

3.5　媒体组件

媒体组件

媒体组件有audio音频组件、image图片组件、video视频组件，audio音频组件用来播放音乐，image图片组件用来显示图片，video视频组件用来播放视频。

3.5.1　audio音频

audio音频组件需要有唯一的id，根据id使用wx.createAudioContext('myAudio')创建音频播放的环境，src属性是音频播放的资源路径，poster属性是音频的播放图片，name属性为音频名称，绑定播放、暂停等事件，具体属性如表3.27所示。

表3.27　audio音频属性

属性	类型	默认值	说明
id	String		video 组件的唯一标识符
src	String		要播放音频的资源地址
loop	Boolean	False	是否循环播放
controls	Boolean	True	是否显示默认控件

续表

属性	类型	默认值	说明
poster	String		默认控件上的音频封面的图片资源地址，如果 controls 属性值为 false，则设置 poster 无效
name	String	未知音频	默认控件上的音频名称，如果 controls 属性值为 false，则设置 name 无效
author	String	未知作者	默认控件上的作者名字，如果 controls 属性值为 false，则设置 author 无效
binderror	EventHandle		当发生错误时触发 error 事件，detail = {errMsg: MediaError.code}
bindplay	EventHandle		当开始/继续播放时触发play事件
bindpause	EventHandle		当暂停播放时触发 pause 事件
bindtimeupdate	EventHandle		当播放进度改变时触发 timeupdate 事件，detail = {currentTime, duration}
bindended	EventHandle		当播放到末尾时触发 ended 事件

MediaError.code错误码如表3.28所示。

表3.28 返回错误码

返回错误码	说明
MEDIA_ERR_ABORTED	获取资源被用户禁止
MEDIA_ERR_NETWORD	网络错误
MEDIA_ERR_DECODE	解码错误
MEDIA_ERR_SRC_NOT_SUPPOERTED	不合适资源

示例代码如下所示：

```
<!-- audio.wxml -->
<audio poster="{{poster}}" name="{{name}}" author="{{author}}" src="{{src}}" id="myAudio" controls loop></audio>

<button type="primary" bindtap="audioPlay">播放</button>
<button type="primary" bindtap="audioPause">暂停</button>
<button type="primary" bindtap="audio14">设置当前播放时间为14秒</button>
<button type="primary" bindtap="audioStart">回到开头</button>
```

代码说明如图3.39所示。

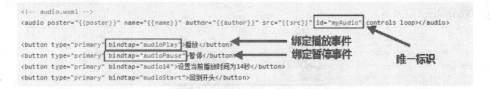

图3.39 audio.wxml代码说明

```
// audio.js
Page({
  onReady: function (e) {
    // 使用 wx.createAudioContext 获取 audio 上下文 context
    this.audioCtx = wx.createAudioContext('myAudio')
```

```
    },
    data: {
        poster: 'http://y.gtimg.cn/music/photo_new/T002R300x300M000003rsKF44GyaSk.jpg?max_age=2592000',
        name: '此时此刻',
        author: '许巍',
        src: 'http://ws.stream.qqmusic.qq.com/M500001VfvsJ21xFqb.mp3?guid=ffffffff82def4af4b12b3cd9337d5e7&uin=346897220&vkey=6292F51E1E384E06DCBDC9AB7C49FD713D632D313AC4858BACB8DDD29067D3C601481D36E62053BF8DFEAF74C0A5CCFADD6471160CAF3E6A&fromtag=46',
    },
    audioPlay: function () {
        this.audioCtx.play()
    },
    audioPause: function () {
        this.audioCtx.pause()
    },
    audio14: function () {
        this.audioCtx.seek(14)
    },
    audioStart: function () {
        this.audioCtx.seek(0)
    }
})
```

代码说明如图3.40所示。

图3.40　audio.js代码说明

界面效果如图3.41所示。

图3.41　音频播放界面效果

3.5.2 image图片

image图片组件的属性如表3.29所示。它有两类展现模式：一类是缩放模式，在缩放模式里包括4种方式；另一类是裁剪模式，在裁剪模式里包括9种方式。

表3.29 image图片属性

属性	类型	默认值	说明
src	String		图片资源地址
mode	String	'scaleToFill'	图片缩放、裁剪的模式
binderror	HandleEvent		当错误发生时，发布到AppService 的事件名，事件对象event.detail = {errMsg: 'something wrong'}
bindload	HandleEvent		当图片载入完毕时，发布到AppService 的事件名，事件对象event.detail = {height:'图片高度px', width:'图片宽度px'}

通过mode属性来设置4种缩放模式，如表3.30所示。

表3.30 4种缩放模式

模式	说明
scaleToFill	不保持纵横比缩放图片，使图片的宽高完全拉伸至填满 image 元素
aspectFit	保持纵横比缩放图片，使图片的长边能完全显示出来。也就是说，可以完整地将图片显示出来
aspectFill	保持纵横比缩放图片，只保证图片的短边能完全显示出来。也就是说，图片通常只在水平或垂直方向是完整的，另一个方向将会发生截取
widthFix	宽度不变，高度自动变化，保持原图宽高比不变

通过mode属性来设置9种裁剪模式，如表3.31所示。

表3.31 9种裁剪模式

模式	说明
top	不缩放图片，只显示图片的顶部区域
bottom	不缩放图片，只显示图片的底部区域
center	不缩放图片，只显示图片的中间区域
left	不缩放图片，只显示图片的左边区域
right	不缩放图片，只显示图片的右边区域
top left	不缩放图片，只显示图片的左上边区域
top right	不缩放图片，只显示图片的右上边区域
bottom left	不缩放图片，只显示图片的左下边区域
bottom right	不缩放图片，只显示图片的右下边区域

示例代码如下所示：

```
<view class="page">
  <view class="page__hd">
    <text class="page__title">image</text>
    <text class="page__desc">图片</text>
  </view>
  <view class="page__bd">
    <view class="section section_gap" wx:for="{{array}}" wx:for-item="item">
      <view class="section__title">{{item.text}}</view>
      <view class="section__ctn">
```

```
            <image style="width: 200px; height: 200px; background-color: #eeeeee;" mode="{{item.mode}}"
src="{{src}}"></image>
        </view>
    </view>
  </view>
</view>

Page({
  data: {
    array: [{
      mode: 'scaleToFill',
      text: 'scaleToFill：不保持纵横比缩放图片，使图片完全适应'
    }, {
      mode: 'aspectFit',
      text: 'aspectFit：保持纵横比缩放图片，使图片的长边能完全显示出来'
    }, {
      mode: 'aspectFill',
      text: 'aspectFill：保持纵横比缩放图片，只保证图片的短边能完全显示出来'
    }, {
      mode: 'top',
      text: 'top：不缩放图片，只显示图片的顶部区域'
    }, {
      mode: 'bottom',
      text: 'bottom：不缩放图片，只显示图片的底部区域'
    }, {
      mode: 'center',
      text: 'center：不缩放图片，只显示图片的中间区域'
    }, {
      mode: 'left',
      text: 'left：不缩放图片，只显示图片的左边区域'
    }, {
      mode: 'right',
      text: 'right：不缩放图片，只显示图片的右边区域'
    }, {
      mode: 'top left',
      text: 'top left：不缩放图片，只显示图片的左上边区域'
    }, {
      mode: 'top right',
      text: 'top right：不缩放图片，只显示图片的右上边区域'
    }, {
      mode: 'bottom left',
      text: 'bottom left：不缩放图片，只显示图片的左下边区域'
    }, {
      mode: 'bottom right',
      text: 'bottom right：不缩放图片，只显示图片的右下边区域'
    }],
    src: '../../resources/cat.jpg'
  },
  imageError: function(e) {
    console.log('image3发生error事件，携带值为', e.detail.errMsg)
  }
})
```

图片效果如图3.42～图3.54所示。

图3.42　原图

图3.43　scaleToFill缩放模式

图3.44　aspectFit缩放模式

图3.45　aspectFill缩放模式

图3.46　top裁剪模式

图3.47　bottom裁剪模式

图3.48　center裁剪模式

图3.49　left裁剪模式

图3.50　right裁剪模式

图3.51　top left裁剪模式

图3.52　top right裁剪模式　　图3.53　bottom left裁剪模式　　图3.54　bottom right裁剪模式

3.5.3　video视频

video视频组件是用来播放视频的组件，这个组件可以控制是否显示默认播放控件（播放/暂停按钮、播放进度、时间），还可以发送弹幕信息等，video组件默认宽度为300px、高度为225px，设置宽高需要通过wxss设置width和height，具体属性如表3.32所示。

表3.32　video视频属性

属性	类型	默认值	说明
src	String		要播放视频的资源地址
controls	Boolean	True	是否显示默认播放控件（播放/暂停按钮、播放进度、时间）
danmu-list	Object	Array	弹幕列表
danmu-btn	Boolean	False	是否显示弹幕按钮，只在初始化时有效，不能动态变更
enable-danmu	Boolean	False	是否展示弹幕，只在初始化时有效，不能动态变更
autoplay	Boolean	False	是否自动播放
bindplay	EventHandle		当开始/继续播放时触发play事件
bindpause	EventHandle		当暂停播放时触发pause事件
bindended	EventHandle		当播放到末尾时触发ended事件
bindtimeupdate	EventHandle		播放进度变化时触发，event.detail = {currentTime: '当前播放时间'}。触发频率应该在 250ms 一次
objectFit	String	Contain	当视频大小与video容器大小不一致时，视频的表现形式。contain：包含。fill：填充。cover：覆盖

示例代码如下所示：

```
<view class="section tc">
  <video id="myVideo" src="http://wxsnsdy.tc.qq.com/105/20210/snsdyvideodownload?filekey=30280201010421301f0201690402534804102ca905ce620b1241b726bc41dcff44e00204012882540400&bizid=1023&hy=SH&fileparam=302c020101042530230204136ffd93020457e3c4ff02024ef202031e8d7f02030f42400204045a320a0201000400" danmu-list="{{danmuList}}" enable-danmu danmu-btn controls></video>
  <view class="btn-area">
    <button bindtap="bindButtonTap">获取视频</button>
    <input bindblur="bindInputBlur"/>
    <button bindtap="bindSendDanmu">发送弹幕</button>
  </view>
```

```
</view>
function getRandomColor () {
  let rgb = []
  for (let i = 0 ; i < 3; ++i){
    let color = Math.floor(Math.random() * 256).toString(16)
    color = color.length == 1 ? '0' + color : color
    rgb.push(color)
  }
  return '#' + rgb.join('')
}

Page({
  onReady: function (res) {
    this.videoContext = wx.createVideoContext('myVideo')
  },
  inputValue: '',
    data: {
        src: '',
      danmuList: [
        {
          text: '第 1s 出现的弹幕',
          color: '#ff0000',
          time: 1
        },
        {
          text: '第 3s 出现的弹幕',
          color: '#ff00ff',
          time: 3
        }]
    },
  bindInputBlur: function(e) {
    this.inputValue = e.detail.value
  },
  bindButtonTap: function() {
    var that = this
    wx.chooseVideo({
      sourceType: ['album', 'camera'],
      maxDuration: 60,
      camera: ['front','back'],
      success: function(res) {
        that.setData({
          src: res.tempFilePath
        })
      }
    })
  },
  bindSendDanmu: function () {
    this.videoContext.sendDanmu({
      text: this.inputValue,
      color: getRandomColor()
    })
```

       }
    })
界面效果如图3.55所示。

图3.55　视频播放界面

3.5.4　camera相机

　　camera相机组件在使用的时候需要用户授权scope.camera，它是由客户端创建的原生组件，它的层级是最高的，不能通过 z-index 控制层级，可使用 cover-view、cover-image覆盖在上面，同一页面只能插入一个 camera 组件，不能在 scroll-view、swiper、picker-view、movable-view 中使用 camera 组件。camera相机组件的属性如表3.33所示。

表3.33　camera相机组件属性

属性	类型	默认值	说明
device-position	String	back	前置或后置，值为front、back
flash	String	Auto	闪光灯，值为auto、on、off
bindstop	EventHandle		摄像头在非正常终止时触发，如退出后台等情况
binderror	EventHandle		用户不允许使用摄像头时触发

示例代码如下所示：

```
<camera device-position="back" flash="off" binderror="error" style="width: 100%; height: 300px;"></camera>
<button type="primary" bindtap="takePhoto">拍照</button>
<view>预览</view>
<image mode="widthFix" src="{{src}}"></image>

Page({
  takePhoto() {
    const ctx = wx.createCameraContext()
    ctx.takePhoto({
      quality: 'high',
      success: (res) => {
        this.setData({
          src: res.tempImagePath
        })
      }
    })
```

```
        },
        error(e) {
            console.log(e.detail)
        }
    })
```

界面效果如图3.56、图3.57所示。

图3.56 拍照前　　　　　　　　　　　图3.57 拍照后

3.5.5　live-player实时音视频播放

　　live-player实时音视频播放组件的使用是针对特定类目开放的，需要先通过类目审核，然后在小程序管理后台"设置"－"接口设置"中自助开通该组件权限。现在支持的类目有：社交（直播）、教育（在线教育）、医疗（互联网医院、公立医院）、政务民生（所有二级类目）、金融（基金、信托、保险、银行、证券/期货、非金融机构自营小额贷款、征信业务、消费金融）。live-player实时音视频播放组件的属性如表3.34所示。

表3.34　live-player实时音视频播放组件属性

属性	类型	默认值	说明
src	String		音视频地址。目前仅支持 flv、rtmp 格式
mode	String	live	live（直播），RTC（实时通话）
autoplay	Boolean	False	自动播放
muted	Boolean	False	是否静音
orientation	String	vertical	画面方向，可选值有 vertical、horizontal
object-fit	String	contain	填充模式，可选值有 contain、fillCrop
background-mute	Boolean	False	进入后台时是否静音（已废弃，默认退台静音）
min-cache	Number	1	最小缓冲区，单位s

续表

属性	类型	默认值	说明
max-cache	Number	3	最大缓冲区，单位s
bindstatechange	EventHandle		播放状态变化事件，detail = {code}
bindfullscreenchange	EventHandle		全屏变化事件，detail = {direction, fullScreen}
bindnetstatus	EventHandle		网络状态通知，detail = {info}

示例代码如下所示：

```
<live-player src="https://domain/pull_stream" mode="RTC" autoplay bindstatechange="statechange" binderror="error" style="width: 300px; height: 225px;" />

Page({
  statechange(e) {
    console.log('live-player code:', e.detail.code)
  },
  error(e) {
    console.error('live-player error:', e.detail.errMsg)
  }
})
```

3.5.6　live-pusher实时音视频录制

　　live-pusher实时音视频录制组件的使用需要取得用户授权scope.camera、scope.record，它是针对特定类目开放的，需要先通过类目审核，然后在小程序管理后台"设置"-"接口设置"中自助开通该组件权限。现在支持的类目有：社交（直播）、教育（在线教育）、医疗（互联网医院、公立医院）、政务民生（所有二级类目）、金融（基金、信托、保险、银行、证券/期货、非金融机构自营小额贷款、征信业务、消费金融）。live-pusher实时音视频录制组件的属性如表3.35所示。

表3.35　live-pusher实时音视频录制组件属性

属性	类型	默认值	说明
url	String	Back	推流地址。目前仅支持 flv、rtmp 格式
mode	String		RTCSD（标清）、HD（高清）、FHD（超清）、RTC（实时通话）
autopush	Boolean	False	自动推流
muted	Boolean	False	是否静音
enable-camera	Boolean	True	开启摄像头
auto-focus	Boolean	True	自动聚集
orientation	String	vertical	vertical, horizontal
beauty	Number	0	美颜
whiteness	Number	0	美白
aspect	String	9:16	宽高比，可选值有3:4、9:16
min-bitrate	Number	200	最小码率
max-bitrate	Number	1000	最大码率
waiting-image	String		进入后台时推流的等待画面

续表

属性	类型	默认值	说明
waiting-image-hash	String		等待画面资源的MD5值
background-mute	Boolean	False	进入后台时是否静音
bindstatechange	EventHandle		状态变化事件，detail = {code}
bindnetstatus	EventHandle		网络状态通知，detail = {info}
binderror	EventHandle		渲染错误事件，detail = {errMsg, errCode}

示例代码如下所示：

```
<live-pusher url="https://domain/push_stream" mode="RTC" autopush bindstatechange="statechange" style="width: 300px; height: 225px;" />

Page({
  statechange(e) {
    console.log('live-pusher code:', e.detail.code)
  }
})
```

3.6 地图组件

地图组件

map地图组件用来开发与地图有关的应用，如地图导航、打车软件、京东商城的订单轨迹等都会用到地图组件，在地图上可以标记覆盖物以及指定一系列的坐标位置。京东的仓库和客户的收货地址，如图3.58所示。

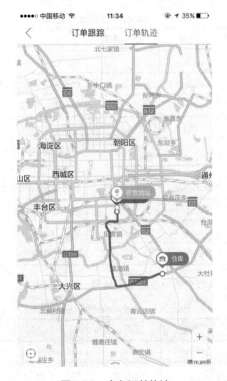

图3.58 京东订单轨迹

map地图组件具体属性如表3.36所示。

表3.36 map地图属性

属性	类型	默认值	说明
longitude	Number		中心经度
latitude	Number		中心纬度
scale	Number	16	缩放级别，取值范围为5~18
markers	Array		标记点
covers	Array		即将移除，请使用 markers
autoplay	Boolean		是否自动播放
polyline	Array		路线
circles	Array		圆
controls	Array		控件
include-points	Array		缩放视野以包含所有给定的坐标点
show-location	Boolean		显示带有方向的当前定位点
bindmarkertap	EventHandle		单击标记点时触发
bindcontroltap	EventHandle		单击控件时触发
bindregionchange	EventHandle		视野发生变化时触发
bindtap	EventHandle		单击地图时触发

markers标记点用于在地图上显示标记的位置，如表3.37所示。

表3.37 markers地图标记属性

属性	说明	类型	是否必填	备注
id	标记点id	Number	否	marker单击事件回调会返回此id
latitude	纬度	Number	是	浮点数，范围为-90~90
longitude	经度	Number	是	浮点数，范围为-180~180
title	标注点名	String	否	
iconPath	显示的图标	String	是	项目目录下的图片路径，支持相对路径写法，以'/'开头则表示相对小程序根目录
rotate	旋转角度	Number	否	顺时针旋转的角度，范围为0~360，默认为 0
alpha	标注的透明度	Number	否	默认为1，无透明
width	标注图标宽度	Number	否	默认为图片实际宽度
height	标注图标高度	Number	否	默认为图片实际高度

polyline指定一系列坐标点，从数组第一项连线至最后一项，如表3.38所示。

表3.38 polyline坐标点属性

属性	说明	类型	是否必填	备注
points	经纬度数组	Array	是	[{latitude: 0, longitude: 0}]
color	线的颜色	String	否	8位十六进制表示，后两位表示alpha值，如：#000000AA

续表

属性	说明	类型	是否必填	备注
width	线的宽度	Number	否	
dottedLine	是否虚线	Boolean	否	默认false

circles在地图上显示圆，如表3.39所示。

表3.39 circles显示圆属性

属性	说明	类型	是否必填	备注
latitude	纬度	Number	是	浮点数，范围为-90~90
longitude	经度	Number	是	浮点数，范围为-180~180
color	描边的颜色	String	否	8位十六进制表示，后两位表示alpha值，如：#000000AA
fillColor	填充颜色	String	否	8位十六进制表示，后两位表示alpha值，如：#000000AA
radius	半径	Number	是	
strokeWidth	描边的宽度	Number	否	

controls在地图上显示控件，控件不随着地图移动，如表3.40所示。

表3.40 controls显示控件属性

属性	说明	类型	是否必填	备注
id	控件id	Number	否	在控件单击事件回调会返回此id
position	控件在地图的位置	Object	是	控件相对地图位置
iconPath	显示的图标	String	是	项目目录下的图片路径，支持相对路径写法，以'/'开头则表示相对小程序根目录
clickable	是否可单击	Boolean	否	默认不可单击

position控件的位置是相对地图的位置，如表3.41所示。

表3.41 position控件位置属性

属性	说明	类型	是否必填	备注
left	距离地图的左边界多远	Number	否	默认为0
top	距离地图的上边界多远	Number	否	默认为0
width	控件宽度	Number	否	默认为图片宽度
height	控件高度	Number	否	默认为图片高度

地图组件的经纬度必须填写，如果不填写经纬度则默认值是北京的经纬度。

示例代码如下所示。

```
<!-- map.wxml -->
```

```html
<map id="map" longitude="113.324520" latitude="23.099994" scale="14" controls="{{controls}}"
bindcontroltap="controltap" markers="{{markers}}" bindmarkertap="markertap" polyline="{{polyline}}"
bindregionchange="regionchange" show-location style="width: 100%; height: 300px;"></map>
```

```js
// map.js
Page({
  data: {
    markers: [{
      iconPath: "/resources/others.png",
      id: 0,
      latitude: 23.099994,
      longitude: 113.324520,
      width: 50,
      height: 50
    }],
    polyline: [{
      points: [{
        longitude: 113.3245211,
        latitude: 23.10229
      }, {
        longitude: 113.324520,
        latitude: 23.21229
      }],
      color:"#FF0000DD",
      width: 2,
      dottedLine: true
    }],
    controls: [{
      id: 1,
      iconPath: '/resources/location.png',
      position: {
        left: 0,
        top: 300 - 50,
        width: 50,
        height: 50
      },
      clickable: true
    }]
  },
  regionchange(e) {
    console.log(e.type)
  },
  markertap(e) {
    console.log(e.markerId)
  },
  controltap(e) {
    console.log(e.controlId)
  }
})
```

界面效果如图3.59所示。

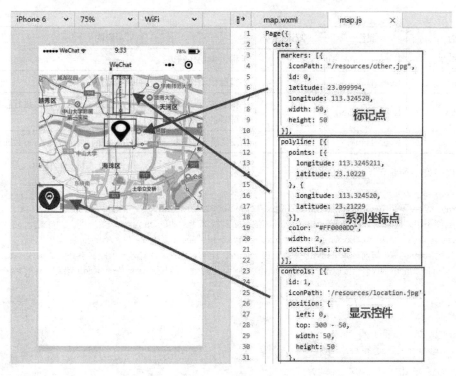

图3.59 地图界面效果

3.7 画布组件

canvas画布组件用来绘制正方形、圆形或者一些其他的形状,如图3.60所示。

画布组件

图3.60 canvas画布组件绘制的图形

canvas画布组件默认宽度为300px、高度为225px,在使用的时候需要有唯一的标识,它有手指触摸动作开始、手指触摸后移动、手指触摸动作结束、手指触摸动作被打断等事件,具体属性如表3.42所示。

表3.42 canvas画布属性

属性	类型	默认值	说明
canvas-id	String		canvas 组件的唯一标识符
disable-scroll	Boolean	False	当在 canvas 中移动时,禁止屏幕滚动以及下拉刷新
bindtouchstart	EventHandle		手指触摸动作开始
bindtouchmove	EventHandle		手指触摸后移动

续表

属性	类型	默认值	说明
bindtouchend	EventHandle		手指触摸动作结束
bindtouchcancel	EventHandle		手指触摸动作被打断，如来电提醒、弹窗
bindlongtap	EventHandle		手指长按 500ms 之后触发，触发了长按事件后进行移动不会触发屏幕的滚动
binderror	EventHandle		当发生错误时触发 error 事件，detail = {errMsg: 'something wrong'}

示例代码如下所示。

```html
<!-- canvas.wxml -->
<canvas style="width: 300px; height: 200px;" canvas-id="firstCanvas"></canvas>
<!-- 当使用绝对定位时，文档流后边的 canvas 的显示层级高于前边的 canvas -->
<canvas style="width: 400px; height: 500px;" canvas-id="secondCanvas"></canvas>
<!-- 因为canvas-id与前一个canvas重复，该canvas不会显示，并会发送一个错误事件到AppService -->
<canvas style="width: 400px; height: 500px;" canvas-id="secondCanvas" binderror="canvasIdErrorCallback"></canvas>
```

```javascript
// canvas.js
Page({
  canvasIdErrorCallback: function (e) {
    console.error(e.detail.errMsg)
  },
  onReady: function (e) {

    // 使用 wx.createContext 获取绘图上下文 context
    var context = wx.createContext()

    context.setStrokeStyle("#00ff00")
    context.setLineWidth(5)
    context.rect(0, 0, 200, 200)
    context.stroke()
    context.setStrokeStyle("#ff0000")
    context.setLineWidth(2)
    context.moveTo(160, 100)
    context.arc(100, 100, 60, 0, 2 * Math.PI, true)
    context.moveTo(140, 100)
    context.arc(100, 100, 40, 0, Math.PI, false)
    context.moveTo(85, 80)
    context.arc(80, 80, 5, 0, 2 * Math.PI, true)
    context.moveTo(125, 80)
    context.arc(120, 80, 5, 0, 2 * Math.PI, true)
    context.stroke()

    // 调用 wx.drawCanvas，通过 canvasId 指定在哪张画布上绘制，通过 actions 指定绘制行为
    wx.drawCanvas({
      canvasId: 'firstCanvas',
      actions: context.getActions() // 获取绘图动作数组
    })
  }
})
```

代码说明如图3.61所示。

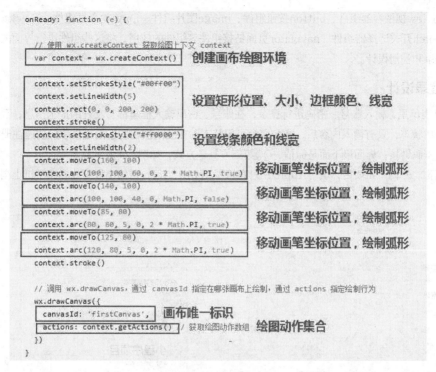

图3.61 绘图代码说明

画布组件以及画布相关API会在后续的章节中深入地学习。

3.8 沙场大练兵：表单登录注册微信小程序

表单登录注册微信小程序

微信小程序里有丰富的表单组件，通过这些组件的使用，来完成京东登录界面、手机快速注册界面、企业用户注册界面的微信小程序设计，如图3.62、图3.63、图3.64所示。

图3.62 登录

图3.63 手机快速注册

图3.64 企业用户注册

会用到view视图容器组件、button按钮组件、image图片组件、input输入框组件、checkbox多项选择器组件、switch开关选择器组件、navigator页面链接组件等组件的使用，将这些组件进行界面的布局设计来完成表单登录和注册设计。

3.8.1 登录设计

在登录表单里，输入账号、密码进行登录，在账号、密码输入框里都有友好的提示信息；登录按钮默认是灰色不可用状态，只有输入内容后，才会变为可用状态；在登录按钮的下面提供手机快速注册、企业用户注册、找回密码链接；界面最下面是微信、QQ第三方登录方式，如图3.65所示。

（1）添加一个form项目，填写AppID，只有填写AppID，form微信小程序才能在手机上浏览效果，如图3.66所示。

图3.65　登录界面

图3.66　添加form项目

（2）在app.json文件里添加"pages/login/login""pages/mobile/mobile""pages/company/company"3个文件目录，并删除默认的文件目录以及相应的文件夹，如图3.67所示。

图3.67　app.json配置

（3）在"pages/login/login"文件里，进行账号密码输入框布局设计，并添加相应的样式，代码如下所示。

```
login.wxml
<view class="content">
  <view class="account">
    <view class="title">账号</view>
    <view class="num"><input bindinput="accountInput" placeholder="用户名/邮箱/手机号" placeholder-style="color:#999999;"/></view>
  </view>
  <view class="hr"></view>
  <view class="account">
    <view class="title">密码</view>
    <view class="num"><input bindblur="pwdBlur" placeholder="请输入密码" password placeholder-style="color:#999999;"/></view>
    <view class="see">
      <image src="/images/see.jpg" style="width:42px;height:30px;"></image>
    </view>
  </view>
<view class="hr"></view>
</view>

login.wxss
.content{
  margin-top: 40px;
}
.account{
  display: flex;
  flex-direction: row;
  padding-left: 20px;
  padding-top: 20px;
  padding-bottom: 10px;
  width: 90%;
}
.title{
  margin-right: 30px;
  font-weight: bold;
}
.hr{
  border: 1px solid #cccccc;
  opacity: 0.2;
  width: 90%;
  margin: 0 auto;
}
.see{
  position: absolute;
  right: 20px;
}
```

界面效果如图3.68所示。

（4）在"pages/login/login"文件里，进行登录按钮、手机快速注册、企业用户注册、找回密码以及第三方登录布局的设计，并添加相应的样式，代码如下所示。

图3.68 输入框布局设计

```
login.wxml
<view class="content">
  <view class="account">
    <view class="title">账号</view>
    <view class="num"><input bindinput="accountInput" placeholder="用户名/邮箱/手机号" placeholder-style="color:#999999;"/></view>
  </view>
  <view class="hr"></view>
  <view class="account">
    <view class="title">密码</view>
    <view class="num"><input bindblur="pwdBlur" placeholder="请输入密码" password placeholder-style="color:#999999;"/></view>
    <view class="see">
      <image src="/images/see.jpg" style="width:42px;height:30px;"></image>
    </view>
  </view>
  <view class="hr"></view>
<button class="btn" disabled="{{disabled}}" type="{{btnstate}}" bindtap="login">登录</button>
<view class="operate">
    <view><navigator url="../mobile/mobile">手机快速注册</navigator></view>
    <view><navigator url="../company/company">企业用户注册</navigator></view>
    <view>找回密码</view>
  </view>
  <view class="login">
    <view><image src="/images/wxlogin.png" style="width:70px;height:98px;"></image></view>
    <view><image src="/images/qqlogin.png" style="width:70px;height:98px;"></image></view>
  </view>
</view>
```

```css
login.wxss
.content{
  margin-top: 40px;
}
.account{
  display: flex;
  flex-direction: row;
  padding-left: 20px;
  padding-top: 20px;
  padding-bottom: 10px;
  width: 90%;
}
.title{
  margin-right: 30px;
  font-weight: bold;
}
.hr{
  border: 1px solid #cccccc;
  opacity: 0.2;
  width: 90%;
  margin: 0 auto;
}
.see{
  position: absolute;
  right: 20px;
}
.btn{
  width: 90%;
  margin-top:40px;
  color: #999999;
}
.operate{
  display: flex;
  flex-direction: row;
}
.operate view{
  margin: 0 auto;
  margin-top:40px;
  font-size: 14px;
  color: #333333;
}
.login{
  display: flex;
  flex-direction: row;
  margin-top:150px;
}
.login view{
  margin: 0 auto;
}
```

界面效果如图3.69所示。

图3.69 布局设计

（5）在"pages/login/login"文件中的js文件里添加accountInput、pwdBlur事件函数，当账号里输入内容后，登录按钮变为可用状态，代码如下所示。

```
login.js
Page({
  data:{
    disabled:true,
    btnstate:"default",
    account:"",
    password:""
  },
  accountInput:function(e){
    var content = e.detail.value;
    console.log(content);
    if(content != ''){
      this.setData({disabled:false,btnstate:"primary",account:content});
    }else{
      this.setData({disabled:true,btnstate:"default"});
    }
  },
  pwdBlur:function(e){
    var password = e.detail.value;
    if(password != ''){
      this.setData({password:password});
    }
  }
})
```

界面效果如图3.70所示。

图3.70 登录按钮可用状态

3.8.2 手机号注册设计

在手机号注册里,需要设计输入框用来输入手机号,设计同意注册协议以及下一步按钮,如图3.71所示。

图3.71 手机号注册界面

(1)在"pages/mobile/mobile"文件里,进行手机号输入框布局设计,并添加相应的样式,代码如下所示。

```
mobile.wxml
<view class="content">
```

```
    <view class="hr"></view>
    <view class="numbg">
        <view>+86</view>
        <view><input placeholder="请输入手机号" maxlength="11" bindblur="mobileblur"/></view>
    </view>
</view>

mobile.wxss
.content{
    width:100%;
    height: 600px;
    background-color: #f2f2f2;
}
.hr{
    padding-top:20px;
}
.numbg{
    border: 1px solid #cccccc;
    width: 90%;
    margin: 0 auto;
    background-color: #ffffff;
    border-radius: 5px;
    display: flex;
    flex-direction: row;
    height: 50px;
}
.numbg view{
    margin-left: 20px;
    margin-top:14px;
}
```

界面效果如图3.72所示。

图3.72　手机号输入框

（2）在"pages/mobile/mobile"文件里，设计注册协议和下一步按钮操作，并添加相应的样式，代码如下所示。

```
mobile.wxml
<view class="content">
    <view class="hr"></view>
    <view class="numbg">
        <view>+86</view>
        <view><input placeholder="请输入手机号" maxlength="11" bindblur="mobileblur"/></view>
    </view>
    <view>
      <view class="xieyi">
          <icon type="success" color="red" size="18"></icon>
          <text class="agree">同意</text>
          <text class="opinion">京东用户注册协议</text>
      </view>
    </view>
    <button class="btn" disabled="{{disabled}}" type="{{btnstate}}" bindtap="login">下一步</button>
</view>
```

```
mobile.wxss
.content{
    width:100%;
    height: 600px;
    background-color: #f2f2f2;
}
.hr{
    padding-top:20px;
}
.numbg{
    border: 1px solid #cccccc;
    width: 90%;
    margin: 0 auto;
    background-color: #ffffff;
    border-radius: 5px;
    display: flex;
    flex-direction: row;
    height: 50px;
}
.numbg view{
    margin-left: 20px;
    margin-top:14px;
}
.xieyi{
    margin-top:15px;
    margin-left:15px;
}
.agree{
    font-size: 13px;
    margin-left: 5px;
    color: #666666;
}
```

```
.opinion{
    font-size: 13px;
    color: #000000;
    font-weight: bold;
}
.btn{
    width:90%;
    margin-top:30px;
}
```

界面效果如图3.73所示。

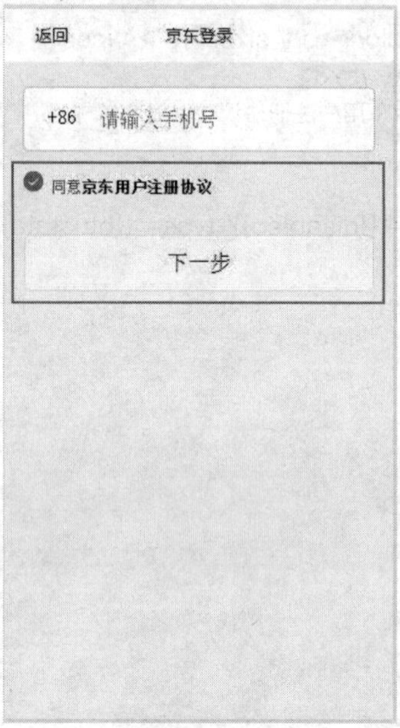

图3.73　同意注册协议及下一步按钮

（3）在"pages/mobile/mobile"文件里，添加mobileblur事件，如果输入手机号，下一步按钮变为可用状态，代码如下所示。

```
mobile.js
Page({
    data:{
        disabled:true,
        btnstate:"default",
        mobile:""
    },
    mobileblur:function(e){
        var content = e.detail.value;
        if(content !=""){
            this.setData({disabled:false,btnstate:"primary",mobile:content});
        }else{
            this.setData({disabled:true,btnstate:"defalut",mobile:""});
        }
    }
})
```

界面效果如图3.74所示。

（4）在mobile.json文件里，添加"navigationBarTitleText"这个属性，设置导航标题为手机快速注册，如图3.75所示。

图3.74　下一步按钮可用

图3.75　导航标题

3.8.3　企业用户注册设计

在企业用户注册里，有6个表单项：用户名、密码、企业全称、联系人姓名、手机号和短信验证码。有一个注册按钮和同意注册协议，如图3.76所示。

图3.76　企业用户注册界面

（1）在"pages/company/company"文件里，进行用户名、密码、企业全称、联系人姓名、手机号、短信验证码表单项布局设计，并添加相应的样式，代码如下所示。

```
company.wxml
<form bindsubmit="formSubmit" bindreset="formReset">
<view class="content">
  <view class="hr"></view>
  <view class="item">
    <input type="text" name="loginName" placeholder="请设置4-20位用户名" placeholder-class="holder" bindblur="accountblur"/>
  </view>
  <view class="item flex">
    <input type="text" password name="password" placeholder="请设置6-20位登录密码" placeholder-class="holder"/>
    <switch type="switch" name="switch"/>
  </view>
  <view class="item">
    <input type="text" name="company" placeholder="请填写工商局注册名称" placeholder-class="holder" />
  </view>
  <view class="item">
    <input type="text" name="userName" placeholder="联系人姓名" placeholder-class="holder" />
  </view>
  <view class="mobileInfo">
    <view class="mobile">
      <input type="text" name="mobile" placeholder="请输入手机号" placeholder-class="holder" />
    </view>
    <view class="code">发送验证码</view>
  </view>
  <view class="item">
    <input type="text" name="code" placeholder="短信验证码" placeholder-class="holder" />
  </view>
</view>
</form>

company.wxss
.content{
    width: 100%;
    height: 700px;
    background-color: #f2f2f2;
}
.hr{
    padding-top:40px;
}
.item{
    margin: 0 auto;
    border: 1px solid #cccccc;
    height: 40px;
    width: 90%;
    border-radius: 3px;
    background-color: #ffffff;
    margin-bottom:15px;
}
```

```css
.item input{
    height: 40px;
    line-height: 40px;
    margin-left: 10px;
}
.holder{
    font-size: 14px;
    color: #999999;
}
.flex{
    display: flex;
    flex-direction: row;
}

.flex input{
    width:300px;
}
.item switch{
    margin-top: 5px;
    margin-right:5px;
}
.mobileInfo{
    display: flex;
    flex-direction: row;
}
.mobile{
    margin: 0 auto;
    border: 1px solid #cccccc;
    height: 40px;
    width: 50%;
    border-radius: 3px;
    background-color: #ffffff;
    margin-bottom:15px;
    display:flex;
    flex-direction: row;
    margin-left:5%;
}
.mobile input{
    margin-top:8px;
    margin-left:10px;
}
.code{
    border: 1px solid #cccccc;
    height: 40px;
    width: 35%;
    background-color: #EFEEEC;
    border-radius:3px;
    text-align: center;
    margin-left:10px;
    line-height:  40px;
    color: #999999;
    font-size: 15px;
```

```
    margin-bottom: 15px;
    margin-right:5%;
}
```
界面效果如图3.77所示。

图3.77 企业用户注册表单项

（2）在"pages/company/company"文件里，设计注册按钮和同意注册协议，并添加相应的样式，代码如下所示。

```
company.wxml
<form bindsubmit="formSubmit" bindreset="formReset">
<view class="content">
  <view class="hr"></view>
  <view class="item">
     <input type="text" name="loginName" placeholder="请设置4-20位用户名" placeholder-class="holder" bindblur="accountblur"/>
  </view>
  <view class="item flex">
     <input type="text" password name="password"  placeholder="请设置6-20位登录密码" placeholder-class="holder"/>
     <switch type="switch" name="switch"/>
  </view>
  <view class="item">
     <input type="text" name="company" placeholder="请填写工商局注册名称" placeholder-class="holder" />
  </view>
  <view class="item">
     <input type="text" name="userName" placeholder="联系人姓名" placeholder-class="holder" />
  </view>
  <view class="mobileInfo">
    <view class="mobile">
      <input type="text" name="mobile" placeholder="请输入手机号" placeholder-class="holder" />
    </view>
```

```
        <view class="code">发送验证码</view>
    </view>
    <view class="item">
        <input type="text" name="code" placeholder="短信验证码" placeholder-class="holder" />
    </view>
    <button class="btn" disabled="{{disabled}}" type="{{btnstate}}" form-type="submit">注册</button>
    <view class="xieyi">
        <text class="agree">注册即视为同意</text><text class="opinion">《京东用户注册协议》</text>
    </view>
</view>
</form>

.content{
    width: 100%;
    height: 700px;
    background-color: #f2f2f2;
}
.hr{
    padding-top:40px;
}
.item{
    margin: 0 auto;
    border: 1px solid #cccccc;
    height: 40px;
    width: 90%;
    border-radius: 3px;
    background-color: #ffffff;
    margin-bottom:15px;
}
.item input{
    height: 40px;
    line-height: 40px;
    margin-left: 10px;
}
.holder{
    font-size: 14px;
    color: #999999;
}
.flex{
    display: flex;
    flex-direction: row;
}
.flex input{
    width:300px;
}
.item switch{
    margin-top: 5px;
    margin-right:5px;
}
.mobileInfo{
    display: flex;
```

```
        flex-direction: row;
}
.mobile{
        margin: 0 auto;
        border: 1px solid #cccccc;
        height: 40px;
        width: 50%;
        border-radius: 3px;
        background-color: #ffffff;
        margin-bottom:15px;
        display:flex;
        flex-direction: row;
        margin-left:5%;
}
.mobile input{
        margin-top:8px;
        margin-left:10px;
}
.code{
        border: 1px solid #cccccc;
        height: 40px;
        width: 35%;
        background-color: #EFEEEC;
        border-radius:3px;
        text-align: center;
        margin-left:10px;
        line-height:   40px;
        color: #999999;
        font-size: 15px;
        margin-bottom: 15px;
        margin-right:5%;
}
.btn{
        width: 90%;
        color: #999999;
        margin-top:40px;
}
.xieyi{
        margin-top:15px;
        margin-left:15px;
        font-size:13px;
}
.agree{
        margin-left: 5px;
        color: #666666;
}
.opinion{
        color: red;
        font-weight: bold;
        text-decoration: underline;
}
```

界面效果如图3.78所示。

第3章 用微信小程序组件构建UI界面

图3.78 注册按钮

（3）当输入用户名后，注册按钮变为可用状态，同时将表单内容提交到company.js文件后台里，保存到缓存里面，代码如下所示。

```
Page({
  data:{
    disabled:true,
    btnstate:"default"
  },
  accountblur:function(e){
    var content = e.detail.value;
    if(content !=""){
      this.setData({disabled:false,btnstate:"primary"});
    }else{
      this.setData({disabled:true,btnstate:"default"});
    }
  },
  formSubmit:function(e){
    console.log(e);
    var user = new Object();
    user.account = e.detail.value.loginName;
    user.password = e.detail.value.password;
    user.company = e.detail.value.company;
    user.userName = e.detail.value.userName;
    user.code = e.detail.value.code;
    user.mobile = e.detail.value.mobile;
    user.switch = e.detail.value.switch;
    wx.setStorageSync('user', user);
    wx.showToast({
        title:"注册成功",
        icon:"success",
```

```
            duration:1000,
            success:function(){
                wx.navigateTo({
                    url: '../login/login'
                })
            }
        });
    }
})
```

(4)在company.json文件里,添加"navigationBarTitleText"这个属性,设置导航标题为企业用户注册。

(5)单击工具栏的上传按钮,将项目上传到微信服务器上,然后单击工具栏上的预览按钮,用微信扫描二维码,可以在手机上预览微信小程序,如图3.79所示。

图3.79 预览微信小程序(本图中二维码只是示意,请扫描自己操作生成的二维码)

这样就完成了京东登录注册表单微信小程序设计,在登录界面进行登录,可以通过手机快速注册进行手机号注册,也可以通过企业用户注册来注册企业账号。

3.9 小结

本章主要学习了微信小程序组件的使用,重点掌握以下内容:

(1)掌握视图容器组件的使用,会制作海报轮播效果、页签切换效果、上下滑动效果以及左右滑动效果;

(2)掌握基础内容组件的使用,包括图标组件、文本组件以及进度条组件的使用;

(3)掌握表单组件的使用,利用表单组件来设计微信小程序的表单内容,可以提交表单以及重置表单内容;

(4)掌握导航组件的使用,保留当前页跳转以及关闭当前页跳转;

(5)掌握媒体组件的使用,包括音频组件、图片组件以及视频组件的使用;

(6)掌握地图组件和画布组件的使用。

第4章
必备的微信小程序API

- 请求服务器数据API
- 文件上传与下载API
- WebSocket会话API
- 图片处理API
- 文件操作API
- 数据缓存API
- 位置信息API
- 设备应用API
- 交互反馈API
- 登录API
- 微信支付API
- 分享API
- 沙场大练兵：仿豆瓣电影微信小程序
- 小结

微信小程序提供了很多在开发微信小程序时会用到的API接口，有请求服务器数据、文件上传与下载、WebSocket会话、图片处理、文件操作、数据缓存、位置信息、设备应用、交互反馈、登录、微信支付、分享等，本章将详细介绍常用的API接口。

4.1 请求服务器数据API

请求服务器数据API

wx.request是用来请求服务器数据的API，它发起的是HTTPS请求，同时它需要在微信公众平台配置HTTPS服务器域名，一个月内可申请修改3次，否则在有AppID创建的项目中无法使用wx.request请求服务器数据这个API，WebSocket会话、文件上传与下载服务器域名都是如此，配置服务器域名如图4.1所示。

图4.1 配置服务器域名

wx.request(object)参数说明如表4.1所示。

表4.1 wx.request参数说明

属性	类型	是否必填	说明
url	String	是	开发者服务器接口地址
Data	Object、String	否	请求的参数
Header	Object	否	设置请求的header，header中不能设置Referer
Method	String	否	默认为GET，有效值为OPTIONS、GET、HEAD、POST、PUT、DELETE、TRACE、CONNECT
dataType	String	否	默认为json。如果设置了dataType为json，则会尝试对响应的数据做一次JSON.parse
Success	Function	否	收到开发者服务器成功返回的回调函数，res = {data: '开发者服务器返回的内容'}
Fail	Function	否	接口调用失败的回调函数
complete	Function	否	接口调用结束的回调函数（调用成功、失败都会执行）

下面演示wx.request请求服务器数据API的使用。

1. 请求HTTP服务器数据

在js文件中的onLoad函数里，使用wx.request请求猫眼电影HTTP服务器数据，具体代码如下所示。

```
Page({
  onLoad:function(){
```

```
wx.request({
    url: 'https://m.maoyan.com/movie/list.json',
    data: {
      type:'hot',
      offset:0,
      limit:1000
    },
    method: 'GET',
    success: function(res){
      console.log(res);
    },
    fail: function() {
      // fail
    },
    complete: function() {
      // complete
    }
  })
}
})
```

请求错误信息如图4.2所示。

图4.2　HTTP请求

从图4.2中可以看出，wx.request无法请求HTTP域名的服务器。访问服务器路径的时候，会到公众开发平台里去找我们配置的HTTPS服务器域名，如果域名存在，就可以访问，否则不可以访问。

data 数据说明最终发送给服务器的数据是 String 类型，如果传入的 data 不是 String 类型，会被转换成String类型。转换规则如下：

对于 header['content-type'] 为 'Application/json' 的数据，会对数据进行 JSON 序列化；

对于 header['content-type'] 为 'Application/x-www-form-urlencoded' 的数据，会将数据转换成 query string （encodeURIComponent(k)=encodeURIComponent(v)&encodeURIComponent(k)=encodeURIComponent(v)...）。

2. 请求HTTPS服务器数据

```
Page({
  onLoad: function () {
    wx.request({
      url: 'https://m.maoyan.com/movie/list.json',
      data: {
        type: "movies",
        offset: 0,
        limit: 10
      },
```

```
      method: "GET",
      header: {
        'content-type': 'application/json'
      },
      success: function (res) {
        console.log(res);
      }
    })
  }
})
```

服务器返回数据如图4.3所示。

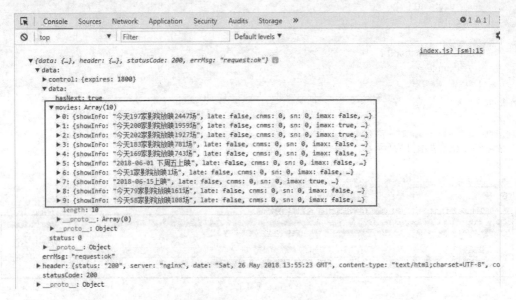

图4.3　服务器返回数据

content-type 默认为 'application/json'，客户端的 HTTPS TLS 版本为1.2，但 Android 的部分机型还未支持 TLS 1.2，所以请确保 HTTPS 服务器的 TLS 版本支持1.2及以下版本；要注意 method 的 value 必须为大写（例如：GET）；url 中不能有端口；request 的默认超时时间和最大超时时间都是 60s，request 的最大并发数是 5，网络请求的 referer 是不可以设置的，格式固定为 https://servicewechat.com/{Appid}/{version}/page-frame.html，其中{Appid} 为小程序的 Appid，{version} 为小程序的版本号，版本号为 0 表示为开发版。

如果项目没有填写AppID，可以访问HTTP请求以及公众开发平台以外的一些服务器请求，但是不填写AppID，在手机上是无法预览和使用的，所以学习过程中可以不填写AppID，来学习这些API的使用。

4.2　文件上传与下载API

文件上传与下载API是我们经常会用到的API，它可以用来与服务器进行文件的上传与下载，比如微信小程序客户端向服务器传输一些图片，或者从服务器那里获得一些图片，这时就可以使用文件上传与下载API，它们请求服务器地址也需要在微信公众平台里进行配置。

文件上传与下载API

4.2.1 wx.uploadFile文件上传

wx.uploadFile(object)参数说明如表4.2所示。

表4.2 wx.uploadFile参数说明

属性	类型	是否必填	说明
url	String	是	开发者服务器 url
filePath	String	是	要上传文件资源的路径
name	String	是	文件对应的 key,开发者在服务器端通过这个 key 可以获取到文件二进制内容
header	Object	否	HTTP 请求 Header,header 中不能设置 Referer
formData	Object	否	HTTP 请求中其他额外的 form data
success	Function	否	接口调用成功的回调函数
fail	Function	否	接口调用失败的回调函数
complete	Function	否	接口调用结束的回调函数(调用成功、失败都会执行)

下面演示一下wx.uploadFile文件上传的使用,将选择的图片传到服务器里。

(1)创建一个无AppID的项目,微信小程序选中上传的图片,利用wx.uploadFile上传图片到服务器,具体代码如下所示。

```
Page({
  onLoad:function(){
    wx.chooseImage({
      count: 9, // 最多可以选择的图片张数,默认为9
      sizeType: ['original', 'compressed'], // original 原图,compressed 压缩图,默认二者都有
      sourceType: ['album', 'camera'], // album 从相册选图,camera 使用相机,默认二者都有
      success: function(res){
        var tempFilePaths = res.tempFilePaths;
        wx.uploadFile({
          url: 'http://localhost:8555/wxApp/WxUploadFileServlet',
          filePath:tempFilePaths[0],
          name:'name',
          header: {
              'content-type': 'Application/json'
          },
          formData: {
              imgName:'我是图片名称',
              imgSize:'122kb'
          },
          success: function(res){
            console.log(res);
          }
        })
      }

    })
  }
})
```

(2)服务器端采用java代码来编写接收文件上传过来的图片信息,将图片保存到服务器上,具体代码如下所示。

```java
package com.xiaogang.App.servlet;

import java.io.File;
import java.io.FileOutputStream;
import java.io.IOException;
import java.io.InputStream;
import java.util.HashMap;
import java.util.List;
import java.util.Map;

import javax.servlet.ServletException;
import javax.servlet.http.HttpServlet;
import javax.servlet.http.HttpServletRequest;
import javax.servlet.http.HttpServletResponse;

import org.apache.commons.fileupload.FileItem;
import org.apache.commons.fileupload.disk.DiskFileItemFactory;
import org.apache.commons.fileupload.servlet.ServletFileUpload;

/**
 *
 * @ClassName: WxUploadFileServlet
 * @Description: 用于接收微信小程序上传的图片
 * @author LG
 * @date 2016年12月5日 下午1:30:14
 *
 */
public class WxUploadFileServlet extends HttpServlet {
    private static final long serialVersionUID = 1L;

    public WxUploadFileServlet() {
        super();
    }

    protected void doGet(HttpServletRequest request, HttpServletResponse response) throws ServletException, IOException {
        doPost(request,response);
    }

    protected void doPost(HttpServletRequest request, HttpServletResponse response) throws ServletException, IOException {
        try {
            DiskFileItemFactory factory = new DiskFileItemFactory();
            ServletFileUpload upload = new ServletFileUpload(factory);
            upload.setHeaderEncoding("UTF-8");
            List items = upload.parseRequest(request);
            Map param = new HashMap();
            for(Object object:items){
                FileItem fileItem = (FileItem) object;
                if (fileItem.isFormField()) {
                    param.put(fileItem.getFieldName(), fileItem.getString("utf-8"));//如果你页面编码是utf-8的
                }else{
                    if("file".equals(fileItem.getFieldName())){
                        String fileName = fileItem.getName();
```

```
                        InputStream is = fileItem.getInputStream();
                        //创建img文件夹
                        new File("E:/img/").mkdir();
                        File file = new File("E:/img/",fileName);
                        file.createNewFile();

                        FileOutputStream fos = new FileOutputStream(file);
                        byte[] b = new byte[1024];
                        while((is.read(b)) != -1){
                            fos.write(b);
                        }
                        is.close();
                        fos.close();
                    }
                }
            }
            System.out.println(param);
        } catch (Exception e) {
            e.printStackTrace();
        }
    }
}
```

（3）文件上传成功后，利用wx.uploadFile里的success回调函数，可以查看文件是否上传成功，如图4.4所示。

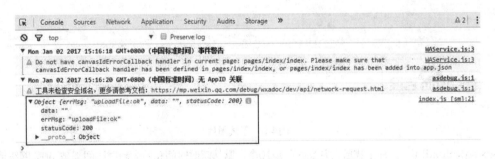

图4.4　回调函数返回值

4.2.2　wx.downloadFile文件下载

wx.downloadFile是文件下载的API，wx.uploadFile是文件上传的API，wx.downloadFile与wx.uploadFile正好相反，是从服务器获得数据，下载到微信小程序客户端本地，参数说明如表4.3所示。

表4.3　wx.downloadFile参数说明

属性	类型	是否必填	说明
url	String	是	开发者服务器 url
header	Object	否	HTTP 请求 Header，header 中不能设置 Referer
success	Function	否	收到开发者服务器成功返回的回调函数，res = {data: '开发者服务器返回的内容'}
fail	Function	否	接口调用失败的回调函数
complete	Function	否	接口调用结束的回调函数（调用成功、失败都会执行）

下面演示一下wx.downloadFile文件下载接口的使用，服务器传递一张图片给微信小程序客户端，将其下载到本地，并显示出来。

（1）在wxml文件里，添加image组件，用来显示服务器传递过来的图片，具体代码如下所示。

```
<image src="{{src}}" style="width:270px;height:126px;"></image>
```

（2）在js文件里，下载一张服务器的图片，将它的临时路径赋值给src，具体代码如下所示。

```
Page({
  data:{
    src:""
  },
  onLoad:function(){
    var page = this;
    wx.downloadFile({
      url: "https://ss0.bdstatic.com/5aV1bjqh_Q23odCf/static/superman/img/logo/bd_logo1_31bdc765.png",
      type: 'image', // 下载资源的类型，用于客户端识别处理，有效值：image/audio/video
      success: function(res){
        console.log(res);
        var tempPath = res.tempFilePath;
        page.setData({src:tempPath});
      }
    })
  }
})
```

界面效果如图4.5所示。

图4.5　下载图片

wx.downloadFile 文件下载最大并发限制是10个，默认超时时间和最大超时时间都是 60s，网络请求的 referer 是不可以设置的，格式固定为 https://servicewechat.com/{Appid}/{version}/page-frame.html，其中{Appid} 为小程序的 Appid，{version} 为小程序的版本号，版本号为 0 表示为开发版。

4.3　WebSocket会话API

WebSocket会话用来创建一个会话连接，创建完会话连接后可以相互通信，像微信聊天和QQ聊天一样，进行通信。它会涉及以下7个API的使用：

（1）wx.connectSocket(OBJECT)创建一个会话连接；

（2）wx.onSocketOpen(CALLBACK)监听WebSocket连接打开事件；

（3）wx.onSocketError(CALLBACK)监听WebSocket错误；

（4）wx.sendSocketMessage(OBJECT)发送数据；

（5）wx.onSocketMessage(CALLBACK) 监听WebSocket接收到服务器的消息事件；

（6）wx.closeSocket()关闭WebSocket连接；

（7）wx.onSocketClose(CALLBACK)监听WebSocket关闭。

WebSocket 会话API

wx.connectSocket(object)参数说明如表4.4所示。

表4.4 wx.connectSocket参数说明

属性	类型	是否必填	说明
url	String	是	开发者服务器 url
data	Object	否	请求的数据
header	Object	否	HTTP 请求 Header，header 中不能设置 Referer
method	String	否	默认是GET，有效值为OPTIONS、GET、HEAD、POST、PUT、DELETE、TRACE、CONNECT
success	Function	否	接口调用成功的回调函数
fail	Function	否	接口调用失败的回调函数
complete	Function	否	接口调用结束的回调函数（调用成功、失败都会执行）

一个微信小程序同时只能有一个 WebSocket 连接，如果当前已存在一个 WebSocket 连接，会自动关闭该连接，并重新创建一个 WebSocket 连接。

wx.sendSocketMessage (object)参数说明如表4.5所示。

表4.5 wx.sendSocketMessage参数说明

属性	类型	是否必填	说明
data	String/ArrayBuffer	否	请求的数据
success	Function	否	接口调用成功的回调函数
fail	Function	否	接口调用失败的回调函数
complete	Function	否	接口调用结束的回调函数（调用成功、失败都会执行）

下面演示WebSocket会话通信的使用。

（1）在wxml文件里进行界面布局设计，创建一个连接按钮，以及发送消息和返回消息的区域，具体代码如下所示。

```
<button type="default" bindtap="createConn">创建连接</button>
<view style="display:flex;flex-direction:row;margin:10px;">
<input type="text" name="msg"  bindblur="getMsg" style="width:200px;border:1px solid #cccccc;"/>
<button type="primary" size="mini" bindtap="send">发送消息</button>
</view>
<view style="height:200px;">
<view style="font-weight:bold;">客户端发送的消息:</view>
<block wx:for="{{sendMsg}}" wx:for-item="item1">
<view style="color:green">{{item1}}</view>
</block>
</view>


<view style="height:200px;">
<view style="font-weight:bold;">服务端返回的消息</view>
<block wx:for="{{resData}}" wx:for-item="item2">
<view style="color:red">{{item2}}</view>
</block>
</view>
```

```
<view style="margin:10px;">{{content}}</view>
<button type="default" bindtap="closeConn">关闭连接</button>
```
界面效果如图4.6所示。

图4.6　WebSocket会话布局

（2）在js文件里，利用WebSocket的API创建一个会话连接、监听连接打开成功和失败、发送消息和接收信息、关闭连接等，具体代码如下所示。

```
Page({
  data:{
    msg:'',
    sendMsg:[],
    socketOpen:false,
    resData:[]
  },
  createConn:function(){
    var page = this;
    wx.connectSocket({
      url: 'ws://localhost:8555/wxApp/getServer',
      data:{
        x: '',
        y: ''
      },
      header:{
        'content-type': 'Application/json'
      },
      method:"GET"
    });
    wx.onSocketOpen(function(res) {
      console.log(res);
```

```javascript
                page.setData({socketOpen:true});
                console.log('WebSocket连接已打开！')
            });
            wx.onSocketError(function(res){
                console.log('WebSocket连接打开失败，请检查！')
            })
        },
        send:function(e){
            if (this.data.socketOpen) {
                console.log(this.data.socketOpen);
                wx.sendSocketMessage({
                    data:this.data.msg
                });
                var sendMsg = this.data.sendMsg;
                sendMsg.push(this.data.msg);
                this.setData({sendMsg:sendMsg});
                var page = this;
                wx.onSocketMessage(function(res) {
                    var resData = page.data.resData;
                    resData.push(res.data);
                    page.setData({resData:resData});
                    console.log(resData);
                    console.log('收到服务器内容：' + res.data)
                })
            } else {
                console.log('WebSocket连接打开失败，请检查！');
            }
        },
        closeConn:function(e){
            wx.closeSocket();
            wx.onSocketClose(function(res) {
                console.log('WebSocket 已关闭！')
            });
        },
        getMsg:function(e){
            var page = this;
            page.setData({msg:e.detail.value});
        }
    })
```

（3）在服务器端，用java语言代码编写WebSocket会话代码，具体代码如下所示。

```java
package com.xiaogang.App.servlet;

import java.io.IOException;

import javax.websocket.OnClose;
import javax.websocket.OnError;
import javax.websocket.OnMessage;
import javax.websocket.OnOpen;
import javax.websocket.RemoteEndpoint;
import javax.websocket.Session;
import javax.websocket.server.ServerEndpoint;
```

```java
@ServerEndpoint("/getServer")
public class WebsocketServer{

    @OnOpen
    public void onOpen(Session session) {
        System.out.println("sessionId="+session.getId());
        final RemoteEndpoint.Basic basic =   session.getBasicRemote();
    try {
            basic.sendText("会话建立成功！！！");
        } catch (IOException e) {
            e.printStackTrace();
        }
    Thread t1=new Thread(new Runnable() {

            @Override
            public void run() {
                try {
                    Thread.currentThread();
                    Thread.sleep(8000);
                    basic.sendText("server get you a msg: what your name?");
                } catch (InterruptedException e) {
                    e.printStackTrace();
                } catch (IOException e) {
                    e.printStackTrace();
                }
            }
        });
        t1.start();
    }
    /**
     * 收到客户端消息时触发
     * @param relationId
     * @param userCode
     * @param message
     * @return
     */
    @OnMessage
    public String onMessage(Session session,String message) {
        System.out.println("pathParams:"+session.getPathParameters());
        System.out.println("requestParams"+session.getRequestParameterMap());
        return    "Got you msg !"+message;
    }

    /**
     * 异常时触发
     * @param relationId
     * @param userCode
     * @param session
     */
    @OnError
```

```java
        public void onError(Throwable throwable,Session session) {
            System.out.println("pathParams:"+session.getPathParameters());
                System.out.println("requestParams"+session.getRequestParameterMap());
            System.out.print("onError"+throwable.toString());
        }

    /**
     * 关闭连接时触发
     * @param relationId
     * @param userCode
     * @param session
     */
    @OnClose
        public void onClose(Session session) {
            System.out.println("pathParams:"+session.getPathParameters());
                System.out.println("requestParams"+session.getRequestParameterMap());
                    System.out.print("onClose ");
        }
}
```

（4）微信小程序客户端和服务器端代码写完之后，就可以单击创建连接按钮，创建一个会话连接，同时可以发送消息给服务端，并接收服务端传递过来的消息，最后关闭连接，如图4.7所示。

图4.7 会话聊天

4.4 图片处理API

图片处理API

微信小程序针对图片处理，提供了4个API：wx.chooseImage(OBJECT)选择图片API、wx.preview Image(OBJECT)预览图片API、wx.getImageInfo(OBJECT)获得图片信息API、 wx.saveImageToPhotosAlbum保存图片到相册API。

4.4.1 wx.chooseImage(OBJECT)选择图片

wx.chooseImage选择图片API可以从本地相册选择图片或使用相机拍照来选择图片，参数说明如表4.6所示。

表4.6 wx. chooseImage参数说明

属性	类型	是否必填	说明
count	Number	否	最多可以选择的图片张数，默认为9张
sizeType	StringArray	否	original 原图，compressed 压缩图，默认二者都有
sourceType	StringArray	否	album 从相册选择图片，camera 使用相机选择图片，默认二者都有
success	Function	否	成功则返回图片的本地文件路径列表 tempFilePaths
fail	Function	否	接口调用失败的回调函数
complete	Function	否	接口调用结束的回调函数（调用成功、失败都会执行）

wx.chooseImage选择图片API，通过count属性可以设置每次最多选择的图片数量，通过sizeType属性来设置显示original原图或者compressed压缩图，通过sourceType属性设置album从相册选图、camera使用相机选图或者两者都可以。

示例代码如下所示。

```
Page({
  onLoad:function(){
    wx.chooseImage({
      count: 9, // 默认为9
      sizeType: ['original', 'compressed'], // 可以指定是原图还是压缩图，默认二者都有
      sourceType: ['album', 'camera'], // 可以指定来源是相册还是相机，默认二者都有
      success: function (res) {
        // 返回选定图片的本地文件路径列表，tempFilePath可以作为img标签的src属性显示图片
        var tempFilePaths = res.tempFilePaths
      }
    })
  }
})
```

4.4.2 wx.previewImage(OBJECT)预览图片

wx.previewImage预览图片API可以用来预览多张图片以及设置默认显示的图片，参数说明如表4.7所示。

表4.7 wx. previewImage参数说明

属性	类型	是否必填	说明
current	String	否	当前显示图片的链接，不填则默认为 urls 的第一张
urls	StringArray	是	需要预览的图片链接列表
success	Function	否	接口调用成功的回调函数
fail	Function	否	接口调用失败的回调函数
complete	Function	否	接口调用结束的回调函数（调用成功、失败都会执行）

示例代码如下所示。

```
Page({
  onLoad:function(){
    wx.previewImage({
      current: 'http://img02.tooopen.com/images/20150928/tooopen_sy_143912755726.jpg', // 当前显示图片的http链接
      urls: [
        "http://img02.tooopen.com/images/20150928/tooopen_sy_143912755726.jpg",
        "http://img06.tooopen.com/images/20160818/tooopen_sy_175866434296.jpg",
        "http://img06.tooopen.com/images/20160818/tooopen_sy_175833047715.jpg"
      ] // 需要预览的图片http链接列表
    })
  }
})
```

界面效果如图4.8、图4.9所示。

图4.8 预览一

图4.9 预览二

4.4.3 wx.getImageInfo(OBJECT)获得图片信息

wx.getImageInfo用来获得图片信息，包括图片的宽度、图片的高度以及图片返回的路径，参数说明如表4.8所示。

表4.8 wx.getImageInfo参数说明

属性	类型	是否必填	说明
src	String	是	图片的路径，可以是相对路径、临时文件路径、存储文件路径、网络图片路径
success	Function	否	接口调用成功的回调函数
fail	Function	否	接口调用失败的回调函数
complete	Function	否	接口调用结束的回调函数（调用成功、失败都会执行）

success返回参数说明如表4.9所示。

表4.9 success参数说明

参数	类型	说明
width	Number	图片宽度，单位为px
height	Number	图片高度，单位为px
path	String	返回图片的本地路径

示例代码如下所示。

```
Page({
  onLoad:function(){
    wx.getImageInfo({
      src: 'http://img02.tooopen.com/images/20150928/tooopen_sy_143912755726.jpg',
      success: function (res) {
        console.log("图片宽度="+res.width);
        console.log("图片高度="+res.height);
        console.log("图片返回路径="+res.path);
      }
    })
  }
})
```

打印信息如图4.10所示。

图4.10 图片信息

4.4.4 wx.saveImageToPhotosAlbum保存图片到相册

微信小程序支持将图片保存到系统相册里，但是前提是需要用户授权，wx.saveImageToPhotosAlbum(OBJECT)参数说明如表4.10所示。

表4.10 wx. saveImageToPhotosAlbum参数说明

属性	类型	是否必填	说明
filePath	StringArray	是	图片文件路径，可以是临时文件路径也可以是永久文件路径
success	Function	否	接口调用成功的回调函数
fail	Function	否	接口调用失败的回调函数
complete	Function	否	接口调用结束的回调函数（调用成功、失败都会执行）

wx.saveImageToPhotosAlbum调用成功后会返回来一个参数：errMsg 调用结果。

示例代码如下所示。

```
Page({
  data: {
    imgUrl: ''
```

```
    },
    onLoad: function () {
      var page = this;
      wx.downloadFile({
        url: "https://ss2.bdstatic.com/70cFvnSh_Q1YnxGkpoWK1HF6hhy/it/u=49292017,22064401&fm=28&gp=0.jpg",
        type: 'image',
        success: function (res) {
          console.log(res);
          var tempPath = res.tempFilePath;

          wx.saveImageToPhotosAlbum({
            filePath: tempPath,
            success:function(res){
              console.log(res);
            }
          })
        }
      })
    }
})
```

4.5 文件操作API

文件操作API

微信小程序针对文件操作提供了5个API：wx.saveFile将文件保存到本地、wx.getSavedFileList获取本地文件列表、wx.getSavedFileInfo获取本地文件信息、wx.getSaved FileInfo删除本地文件、wx.openDocument打开文档。

4.5.1 wx.saveFile保存文件到本地

wx.saveFile(object)可以将文件保存到本地，下次启动微信小程序的时候，仍然可以获取到该文件。如果是临时路径，下次启动微信小程序的时候，就无法获取到该文件，本地文件存储的大小限制为10MB。参数说明如表4.11所示。

表4.11 wx. saveFile参数说明

属性	类型	必填	说明
tempFilePath	String	是	需要保存的文件的临时路径
success	Function	否	返回文件的保存路径，res = {savedFilePath: '文件的保存路径'}
fail	Function	否	接口调用失败的回调函数
complete	Function	否	接口调用结束的回调函数（调用成功、失败都会执行）

示例代码如下所示。

```
Page({
  onLoad:function(){
    wx.getImageInfo({
      src: 'https://ss2.bdstatic.com/70cFvnSh_Q1YnxGkpoWK1HF6hhy/it/u=49292017,22064401&fm=28&gp=0.jpg',
      success: function (res) {
        var path = res.path;
        console.log("临时文件路径="+path);
        wx.saveFile({
```

```
            tempFilePath: path,
            success: function(res){
                var savedFilePath = res.savedFilePath;
                console.log("本地文件路径="+savedFilePath);
            }
        })
     }
   })
  }
})
```

将文件保存到本地后，会返回一个savedFilePath本地文件存储路径，根据这个路径就可以访问或者使用该文件，同时在下次启动微信小程序的时候，这个本地文件仍然存在。

4.5.2 wx.getSavedFileList获取本地文件列表

通过wx.saveFile可以将临时文件保存到本地，成为本地文件。可以用wx.getSavedFileList来获取本地文件列表，可以获取到wx.saveFile保存的文件，参数说明如表4.12所示。

表4.12　wx. getSavedFileList参数说明

属性	类型	是否必填	说明
success	Function	否	接口调用成功的回调函数，返回结果见success返回参数说明
fail	Function	否	接口调用失败的回调函数
complete	Function	否	接口调用结束的回调函数（调用成功、失败都会执行）

success返回参数说明如表4.13所示。

表4.13　success参数说明

参数	类型	说明
errMsg	String	接口调用结果
fileList	Object Array	文件列表

fileList中的项目说明如表4.14所示。

表4.14　fileList说明

键	类型	说明
filePath	String	文件的本地路径
createTime	Number	文件保存时的时间戳，从1970/01/01 08:00:00 到当前时间的秒数
size	Number	文件大小，单位为B

示例代码如下所示。

```
Page({
  onLoad:function(){
    wx.getSavedFileList({
      success: function(res) {
        var fileList = res.fileList;
        console.log(fileList)
        for(var i=0;i<fileList.length;i++){
          var file = fileList[i];
```

```
                    console.log("第"+(i+1)+"个文件:");
                    console.log("文件创建时间="+file.createTime);
                    console.log("文件大小="+file.size);
                    console.log("文件本地路径="+file.filePath);
                }
            }
        })
    }
})
```

打印信息如图4.11所示。

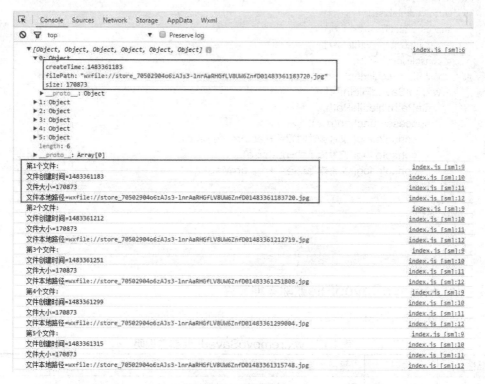

图4.11 本地文件信息

4.5.3 wx.getSavedFileInfo获取本地文件信息

wx.getSavedFileInfo可以获取本地指定路径的文件信息，包括文件的创建时间、文件大小以及接口调用结果，wx.getSavedFileInfo参数说明如表4.15所示。

表4.15 wx. getSavedFileInfo参数说明

属性	类型	是否必填	说明
filePath	String	是	文件路径
success	Function	否	接口调用成功的回调函数，返回结果见success返回参数说明
fail	Function	否	接口调用失败的回调函数
complete	Function	否	接口调用结束的回调函数（调用成功、失败都会执行）

success返回参数说明如表4.16所示。

表4.16 success参数说明

参数	类型	说明
errMsg	String	接口调用结果
size	Number	文件大小，单位为B
createTime	Number	文件保存时的时间戳，从1970/01/01 08:00:00 到当前时间的秒数

示例代码如下所示。

```
Page({
  onLoad:function(){
    wx.getSavedFileList({
      success: function(res) {
        var fileList = res.fileList;
        console.log(fileList)
        var file = fileList[0];
        wx.getSavedFileInfo({
          filePath: file.filePath,
          success: function(res){
            console.log("文件创建时间="+res.createTime);
            console.log("文件大小="+res.size);
            console.log("文件本地路径="+res.errMsg);
          }
        })
      }
    })
  }
})
```

4.5.4 wx.removeSavedFile删除本地文件

wx.saveFile用来将文件保存到本地，而wx.removeSavedFile用来删除本地文件，参数说明如表4.17所示。

表4.17 wx.removeSavedFile参数说明

属性	类型	是否必填	说明
filePath	String	是	需要删除的文件路径
success	Function	否	接口调用成功的回调函数
fail	Function	否	接口调用失败的回调函数
complete	Function	否	接口调用结束的回调函数（调用成功、失败都会执行）

示例代码如下所示。

```
Page({
  onLoad:function(){
    wx.getSavedFileList({
      success: function(res) {
        var fileList = res.fileList;
        console.log(fileList)
        var file = fileList[0];
        wx.removeSavedFile({
          filePath: file.filePath,
          complete: function(res) {
```

```
            console.log(res)
          }
        })
      }
    })
  }
})
```

4.5.5　wx.openDocument打开文档

wx.openDocument可以打开doc、xls、ppt、pdf、docx、xlsx、pptx等多种格式的文档，参数说明如表4.18所示。

表4.18　wx.openDocument参数说明

属性	类型	是否必填	说明
filePath	String	是	文件路径，可通过 downFile 获得
success	Function	否	接口调用成功的回调函数
fail	Function	否	接口调用失败的回调函数
complete	Function	否	接口调用结束的回调函数（调用成功、失败都会执行）

示例代码如下所示。

```
Page({
  onLoad:function(){
    wx.downloadFile({
      url: 'http://www.crcc.cn/portals/0/word/应聘材料样本.doc',
      success: function (res) {
        var filePath = res.tempFilePath
        wx.openDocument({
          filePath: filePath,
          success: function (res) {
            console.log('打开文档成功')
          }
        })
      }
    })
  }
})
```

4.5.6　wx.getFileInfo获取文件信息

wx.getFileInfo可以用来获取文件的大小和摘要信息等内容，参数说明如表4.19所示。

表4.19　wx.getFileInfo参数说明

属性	类型	是否必填	说明
filePath	String	是	本地文件路径
digestAlgorithm	String	否	计算文件摘要的算法，默认值为md5，有效值为md5、sha1
fail	Function	否	接口调用失败的回调函数
complete	Function	否	接口调用结束的回调函数（调用成功、失败都会执行）
success	Function	否	接口调用成功的回调函数

示例代码如下所示：

```
Page({
  onLoad: function () {
    wx.getFileInfo({
      filePath:'https://ss2.bdstatic.com/70cFvnSh_Q1YnxGkpoWK1HF6hhy/it/u=49292017,22064401&fm=28&gp=0.jpg',
      complete:function(res){
        console.log(res.size)
        console.log(res.digest)
      }
    })
  }
})
```

4.6 数据缓存API

数据缓存API

微信小程序数据缓存API用来处理数据缓存信息，可以将数据缓存到本地、获取到本地缓存数据、移除缓存数据以及清理缓存数据，常用的数据缓存API有：

（1）wx.setStorage(OBJECT) 异步方式将数据存储在本地缓存中指定的 key 中；

（2）wx.setStorageSync(KEY,DATA)同步方式将数据存储在本地缓存中指定的key中；

（3）wx.getStorage(OBJECT)异步方式从本地缓存中异步获取指定 key 对应的内容；

（4）wx.getStorageSync(KEY)同步方式从本地缓存中同步获取指定 key 对应的内容；

（5）wx.getStorageInfo(OBJECT)异步方式获取当前storage的相关信息；

（6）wx.getStorageInfoSync同步方式获取当前storage的相关信息；

（7）wx.removeStorage(OBJECT) 异步方式从本地缓存中移除指定key；

（8）wx.removeStorageSync(KEY) 同步方式从本地缓存中移除指定key；

（9）wx.clearStorage()异步方式清理本地数据缓存；

（10）wx.clearStorageSync()同步方式清理本地数据缓存。

4.6.1 数据缓存到本地

微信小程序为数据缓存到本地提供了两种方式，一种是wx.setStorage(OBJECT)异步方式将数据存储在本地缓存中指定的 key 中，另一种是wx.setStorageSync(KEY,DATA)同步方式将数据存储在本地缓存中指定的 key 中，本地缓存最大为10MB。

1. wx.setStorage(OBJECT)

异步方式将数据存储在本地缓存中指定的 key 中，会覆盖掉原来该 key 对应的内容，参数说明如表4.20所示。

表4.20　wx. setStorage参数说明

属性	类型	是否必填	说明
key	String	是	本地缓存中指定的 key
data	Object/String	是	需要存储的内容
success	Function	否	接口调用成功的回调函数
fail	Function	否	接口调用失败的回调函数
complete	Function	否	接口调用结束的回调函数（调用成功、失败都会执行）

如果我们想把用户信息缓存到本地，示例代码如下所示。

```
Page({
  onLoad:function(){
    var user = this.getUserInfo();
    console.log(user);
    wx.setStorage({
      key: 'user',
      data: user,
      success: function(res){
        console.log(res);
      }
    })
  },
  getUserInfo:function(){
    var user = new Object();
    user.name = 'xiaogang';
    user.sex = '男';
    user.age = 30;
    user.address='北京市';
    return user;
  }
})
```

在Storage里可以查看缓存的数据，如图4.12所示。

图4.12　本地缓存数据

2．wx.setStorageSync(KEY,DATA)

同步方式将数据存储到本地指定的key中，会覆盖掉原来该 key 对应的内容，相比于异步缓存数据，它更简单一些，参数说明如表4.21所示。

表4.21　wx.setStorageSync参数说明

属性	类型	是否必填	说明
key	String	是	本地缓存中指定的key
data	Object/String	是	需要存储的内容

示例代码如下所示。

```
Page({
  onLoad:function(){
    var userSync = this.getUserInfo();
    //同步方式将数据存储到本地
    wx.setStorageSync('userSync', userSync)
  },
  getUserInfo:function(){
    var user = new Object();
    user.name = 'xiaogang';
    user.sex = '男';
```

```
        user.age = 30;
        user.address='北京市';
        return user;
    }
})
```

在Storage里可以查看缓存的数据，如图4.13所示。

图4.13 本地缓存数据

数据缓存到本地，不管是同步方式还是异步方式，都是通过key/value的形式存储数据的。

4.6.2 获取本地缓存数据

获取本地缓存数据提供了4个API接口：wx.getStorage(OBJECT)异步方式从本地缓存中异步获取指定 key 对应的内容；wx.getStorageSync(KEY)同步方式从本地缓存中同步获取指定 key 对应的内容；wx.getStorageInfo(OBJECT)异步方式获取当前storage的相关信息；wx.getStorageInfoSync同步方式获取当前storage的相关信息。前两个API接口通过指定key值获得缓存数据，而后两个API接口是获取当前storage相关信息。

1. wx. getStorage (OBJECT)

wx.getStorage(OBJECT)使用异步方式从本地缓存中获取指定 key 对应的内容。参数说明如表4.22所示。

表4.22 wx. getStorage参数说明

属性	类型	是否必填	说明
key	String	是	本地缓存中指定的 key
success	Function	否	接口调用的回调函数,res = {data: key对应的内容}
fail	Function	否	接口调用失败的回调函数
complete	Function	否	接口调用结束的回调函数（调用成功、失败都会执行）

在4.6.1中，使用wx.setStorage(OBJECT)，将user异步方式保存到本地，下面使用wx.getStorage(OBJECT)来获取本地数据，具体代码如下所示。

```
Page({
    onLoad:function(){
        //异步方式获取本地数据
        wx.getStorage({
            key: 'user',
            success: function(res){
                console.log(res);
            }
        })
    }
})
```

获取到的本地数据如图4.14所示。

图4.14 异步获取本地数据

2. wx.getStorageSync (OBJECT)

wx.getStorageSync (OBJECT)是一个同步的接口，用来从本地缓存中同步获取指定 key 对应的内容，它只有一个参数，如表4.23所示。

表4.23 wx.getStorageSync参数说明

属性	类型	是否必填	说明
key	String	是	本地缓存中指定的 key

示例代码如下所示。

```
Page({
  onLoad:function(){
    //同步方式获取本地数据
    var userSync = wx.getStorageSync('userSync');
    console.log(userSync);
  }
})
```

获取到的本地数据如图4.15所示。

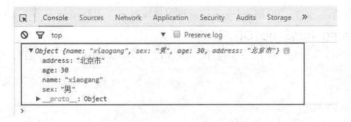

图4.15 同步获取本地数据

3. wx.getStorageInfo (OBJECT)

wx.getStorage和wx.getStorageSync这两个接口都是从本地指定key值来获取数据，wx.getStorageInfo用异步方式获取当前storage的相关信息，来获取所有key的值，参数说明如表4.24所示。

表4.24 wx.getStorageInfo参数说明

属性	类型	是否必填	说明
success	Function	否	接口调用成功的回调函数，详见返回参数说明
fail	Function	否	接口调用失败的回调函数
complete	Function	否	接口调用结束的回调函数（调用成功、失败都会执行）

success返回参数说明如表4.25所示。

表4.25 success参数说明

参数	类型	说明
keys	String Array	当前storage中所有的key
currentSize	Number	当前占用的空间大小，单位为kb
limitSize	Number	限制的空间大小，单位为kb

示例代码如下所示。

```
Page({
  onLoad:function(){
    wx.getStorageInfo({
      success: function(res){
        console.log(res);
      }
    })
  }
})
```

返回值信息如图4.16所示。

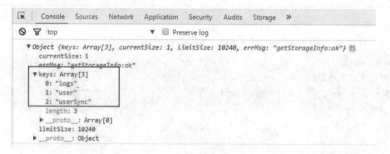

图4.16 异步获取本地key数据

获取到本地所有的key值后，根据这个key值就可以查找到相应的数据。

4. wx. getStorageInfoSync (OBJECT)

wx.getStorageInfoSync是一个用同步的方法来获取当前storage相关信息的API，示例代码如下所示。

```
Page({
  onLoad:function(){
    var storage = wx.getStorageInfoSync();
    console.log(storage);
  }
})
```

返回值信息如图4.17所示。

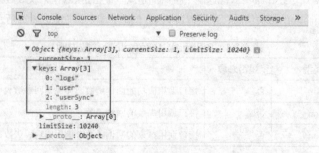

图4.17 同步获取本地key数据

它和wx.getStorageInfo异步获取storage返回数据一样，都是返回所有的key值，然后根据key值再查找完整的数据。

4.6.3 移除和清理本地缓存数据

wx.removeStorage(OBJECT)、wx.removeStorageSync(KEY)用来从本地缓存中移除指定 key；wx.clearStorage()、wx.clearStorageSync()用来清理本地数据缓存。

1. wx.removeStorage(OBJECT)

wx.removeStorage(OBJECT)用异步方式从本地缓存中移除指定的key，参数说明如表4.26所示。

表4.26 wx.removeStorage参数说明

属性	类型	是否必填	说明
key	String	是	本地缓存中指定的 key
success	Function	否	接口调用成功的回调函数
fail	Function	否	接口调用失败的回调函数
complete	Function	否	接口调用结束的回调函数（调用成功、失败都会执行）

在未清理前可以看到本地缓存中有key=user数据，如图4.18所示。

图4.18 user缓存数据

下面从本地缓存中清理key=user的数据，具体代码如下所示。

```
Page({
  onLoad:function(){
    //异步移除key=user数据
    wx.removeStorage({
      key: 'user',
      success: function(res){
        console.log(res);
      },
    })
  }
})
```

清理完成后，不存在key=user的数据，如图4.19所示。

图4.19 清理user缓存数据

2. wx. removeStorageSync(KEY)

wx. removeStorageSync (OBJECT)用来同步从本地缓存中移除指定的key，它的效果和wx. removeStorage一样，参数说明如表4.27所示。

表4.27　wx. removeStorageSync参数说明

属性	类型	是否必填	说明
key	String	是	本地缓存中指定的 key

示例代码如下所示。

```
Page({
  onLoad:function(){
    //同步移除key=userSync数据
    wx.removeStorageSync('userSync');
  }
})
```

3. wx. clearStorage ()、wx. clearStorageSync ()

wx. clearStorage ()、wx. clearStorageSync ()用来清理本地所有缓存数据，wx. clearStorage ()是异步清理缓存数据，wx. clearStorageSync ()是同步清理缓存数据。

示例代码如下所示。

```
wx.clearStorage()

try {
    wx.clearStorageSync()
} catch(e) {
}
```

4.7　位置信息API

微信小程序针对位置新提供了4个API接口：wx.getLocation(OBJECT)获得当前位置信息、wx.chooseLocation(OBJECT)打开地图选择位置、wx.openLocation(OBJECT)使用微信内置地图打开位置和wx.createMapContext(mapId)地图组件控制创建并返回map 上下文 mapContext 对象。

位置信息API

4.7.1　获得位置、选择位置、打开位置

1. wx.getLocation(OBJECT)获得当前位置

使用wx.getLocation(OBJECT)可以获得当前位置信息，包括当前位置的地理坐标、速度，用户离开小程序后，此接口无法调用；当用户点击"显示在聊天顶部"时，此接口可继续调用。具体参数如表4.28所示。

表4.28　wx. getLocation参数说明

属性	类型	是否必填	说明
type	String	否	默认为 wgs84 返回 gps 坐标，gcj02 返回可用于wx.openLocation的坐标
success	Function	是	接口调用成功的回调函数，详见返回参数说明
fail	Function	否	接口调用失败的回调函数
complete	Function	否	接口调用结束的回调函数（调用成功、失败都会执行）

success返回参数说明如表4.29所示。

表4.29 success参数说明

参数	说明
latitude	纬度，浮点数，范围为-90~90，负数表示南纬
longitude	经度，浮点数，范围为-180~180，负数表示西经
speed	速度，浮点数，单位m/s
accuracy	位置的精确度

示例代码如下所示。

```
Page({
  onLoad:function(){
    wx.getLocation({
      type: 'wgs84',
      success: function(res) {
        var latitude = res.latitude;
        console.log("纬度="+latitude);
        var longitude = res.longitude;
        console.log("经度="+longitude);
        var speed = res.speed;
        console.log("速度="+speed);
        var accuracy = res.accuracy;
        console.log("精确度="+accuracy);
      }
    })
  }
})
```

2. wx.chooseLocation(OBJECT)选择位置

使用wx.chooseLocation打开地图来选择位置，具体参数如表4.30所示。

表4.30 wx.chooseLocation参数说明

属性	类型	是否必填	说明
success	Function	是	接口调用成功的回调函数，详见返回参数说明
cancel	Function	否	用户取消时调用
fail	Function	否	接口调用失败的回调函数
complete	Function	否	接口调用结束的回调函数（调用成功、失败都会执行）

success返回参数说明如表4.31所示。

表4.31 success参数说明

参数	说明
latitude	纬度，浮点数，范围为-90~90，负数表示南纬
longitude	经度，浮点数，范围为-180~180，负数表示西经
name	位置名称
address	详细地址

示例代码如下所示。

```
Page({
  onLoad:function(){
    wx.chooseLocation({
      success: function(res){
        console.log(res);
      }
    })
  }
})
```

在地图中选择位置，如图4.20所示。

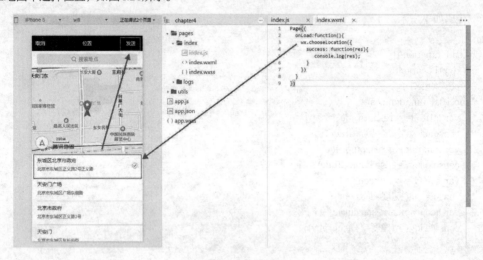

图4.20 选择位置

单击发送，就可以把选择的位置信息打印出来，如图4.21所示。

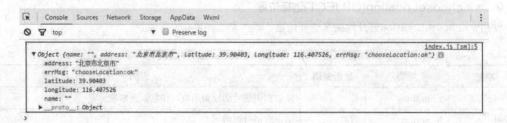

图4.21 位置信息

3. wx.openLocation(OBJECT)打开位置

使用wx.openLocation(OBJECT)这个接口可以使用微信内置地图查看位置，具体参数如表4.32所示。

表4.32 wx. openLocation参数说明

属性	类型	是否必填	说明
latitude	Float	是	纬度，范围为-90～90，负数表示南纬
longitude	Float	是	经度，范围为-180～180，负数表示西经
scale	INT	否	缩放比例，范围为5～18，默认为18
name	String	否	位置名
address	String	否	地址的详细说明

续表

属性	类型	是否必填	说明
success	Function	是	接口调用成功的回调函数
fail	Function	否	接口调用失败的回调函数
complete	Function	否	接口调用结束的回调函数（调用成功、失败都会执行）

示例代码如下所示。

```
Page({
  onLoad:function(){
    wx.getLocation({
      type: 'gcj02', //返回可以用于wx.openLocation的经纬度
      success: function(res) {
        var latitude = res.latitude
        var longitude = res.longitude
        wx.openLocation({
          latitude: latitude,
          longitude: longitude,
          scale: 28
        })
      }
    })
  }
})
```

界面效果如图4.22所示。

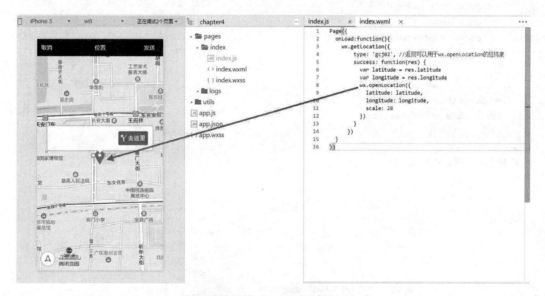

图4.22　打开位置

4.7.2　地图组件控制

wx.createMapContext(mapId)地图组件控制是用来创建并返回 map 上下文 mapContext 对象的，它有两种方法：一种是getCenterLocation，获取当前地图中心的经纬度，返回的是GCJ-02坐标系，可以用于wx.openLocation；另一种是moveToLocation，将地图中心移动到当前定位点，需要配合map组件的showlocation使用。

getCenterLocation方法具体参数如表4.33所示。

表4.33 getCenterLocation参数说明

属性	类型	是否必填	说明
success	Function	是	接口调用成功的回调函数，res = { longitude: "经度", latitude: "纬度"}
fail	Function	否	接口调用失败的回调函数
complete	Function	否	接口调用结束的回调函数（调用成功、失败都会执行）

示例代码如下所示。

```
<!-- map.wxml -->
<map id="myMap" show-location />

<button type="primary" bindtap="getCenterLocation">获取位置</button>
<button type="primary" bindtap="moveToLocation">移动位置</button>

// map.js
Page({
  onReady: function (e) {
    // 使用 wx.createMapContext 获取 map 上下文
    this.mapCtx = wx.createMapContext('myMap')
  },
  getCenterLocation: function () {
    this.mapCtx.getCenterLocation({
      success: function(res){
        console.log(res.longitude)
        console.log(res.latitude)
      }
    })
  },
  moveToLocation: function () {
    this.mapCtx.moveToLocation()
  }
})
```

界面效果如图4.23所示。

图4.23 地图组件控制界面效果

4.8 设备应用API

微信小程序针对设备应用有6类API：获得系统信息、获取网络状态、加速度计、罗盘、拨打电话、扫码。这6类都是针对设备应用提供的相关接口。

设备应用API

4.8.1 获得系统信息

获得系统信息提供了两个API：一个是异步获取系统信息的wx.getSystemInfo(OBJECT)，另一个是同步获取系统信息的wx.getSystemInfoSync()。

1. wx.getSystemInfo(OBJECT)异步获取系统信息

wx.getSystemInfo(OBJECT)用来异步获取设备的系统信息，具体参数如表4.34所示。

表4.34　wx. getSystemInfo参数说明

属性	类型	是否必填	说明
success	Function	是	接口调用成功的回调函数
fail	Function	否	接口调用失败的回调函数
complete	Function	否	接口调用结束的回调函数（调用成功、失败都会执行）

success返回参数说明如表4.35所示。

表4.35　success参数说明

参数	说明
model	手机型号
pixelRatio	设备像素比
windowWidth	窗口宽度
windowHeight	窗口高度
language	微信设置的语言
version	微信版本号
system	操作系统版本
platform	客户端平台

示例代码如下所示。

```
Page({
  onLoad:function(){
    wx.getSystemInfo({
      success: function(res) {
        console.log("手机型号="+res.model)
        console.log("设备像素比="+res.pixelRatio)
        console.log("窗口宽度="+res.windowWidth)
        console.log("窗口高度="+res.windowHeight)
        console.log("微信设置的语言="+res.language)
        console.log("微信版本号="+res.version)
        console.log("操作系统版本="+res.system)
        console.log("客户端平台="+res.platform)
      }
    })
```

    }
  })

2. wx.getSystemInfoSync同步获取系统信息

wx.getSystemInfoSync同步获取系统信息是没有参数的，示例代码如下所示。

```
Page({
  onLoad: function () {
    try {
      var res = wx.getSystemInfoSync()
      console.log("手机型号=" + res.model)
      console.log("设备像素比=" + res.pixelRatio)
      console.log("窗口宽度=" + res.windowWidth)
      console.log("窗口高度=" + res.windowHeight)
      console.log("微信设置的语言=" + res.language)
      console.log("微信版本号=" + res.version)
      console.log("操作系统版本=" + res.system)
      console.log("客户端平台=" + res.platform)
    } catch (e) {
      // Do something when catch error
    }
  }
})
```

4.8.2 获取网络状态

微信小程序使用wx.getNetworkType(OBJECT)来获取网络类型，网络类型分为2G、3G、4G、Wi-Fi，具体参数如表4.36所示。

表4.36　wx.getNetworkType参数说明

属性	类型	是否必填	说明
success	Function	是	接口调用成功，返回网络类型 networkType
fail	Function	否	接口调用失败的回调函数
complete	Function	否	接口调用结束的回调函数（调用成功、失败都会执行）

示例代码如下所示。

```
Page({
  onLoad: function () {
    wx.getNetworkType({
      success: function (res) {
        // 返回网络类型2G、3G、4G、Wi-Fi
        var networkType = res.networkType;
        console.log("网络类型="+networkType);
      }
    })
  }
})
```

4.8.3 加速度计

1. wx.onAccelerometerChange(CALLBACK)监听加速度数据

微信小程序使用wx.onAccelerometerChange(CALLBACK)来进行加速度数据监听，监听加速度计数据，频率为5次/秒，具体参数如表4.37所示。

表4.37　wx. onAccelerometerChange参数说明

参数	类型	说明
X	Number	X轴
Y	Number	Y轴
Z	Number	Z轴

示例代码如下所示。

```
Page({
  onLoad: function () {
    wx.onAccelerometerChange(function(res) {
      console.log("X轴="+res.x)
      console.log("Y轴="+res.y)
      console.log("Z轴="+res.z)
    })
  }
})
```

2. wx.startAccelerometer(OBJECT)开始监听加速度数据

微信小程序使用wx.startAccelerometer(OBJECT)来开始监听加速度数据，具体参数如表4.38所示。

表4.38　wx. startAccelerometer参数说明

属性	类型	是否必填	说明
success	Function	是	接口调用成功的回调函数
fail	Function	否	接口调用失败的回调函数
complete	Function	否	接口调用结束的回调函数（调用成功、失败都会执行）

示例代码如下所示。

```
wx.startAccelerometer()
```

3. wx.stopAccelerometer(OBJECT)停止监听加速度数据

微信小程序使用wx.stopAccelerometer(OBJECT)来停止监听加速度数据，具体参数如表4.39所示。

表4.39　wx. stopAccelerometer参数说明

属性	类型	是否必填	说明
success	Function	是	接口调用成功的回调函数
fail	Function	否	接口调用失败的回调函数
complete	Function	否	接口调用结束的回调函数（调用成功、失败都会执行）

示例代码如下所示。

```
wx.stopAccelerometer()
```

4.8.4　罗盘

微信小程序使用wx.onCompassChange(CALLBACK)来监听罗盘数据，频率为5次/秒，具体参数如表4.40所示。

表4.40　wx. onCompassChange参数说明

参数	类型	说明
direction	Number	面对的方向度数

示例代码如下所示。

```
Page({
  onLoad: function () {
    wx.onCompassChange(function (res) {
      console.log("面对的方向度数="+res.direction)
    })
  }
})
```

微信小程序开始监听罗盘数据使用wx.startCompass(OBJECT), 停止监听罗盘数据使用wx.stopCompass(OBJECT)。

示例代码如下所示。

```
Page({
  onLoad: function () {
    wx.startCompass() //开始监听罗盘数据
    wx.onCompassChange(function (res) {
      console.log("面对的方向度数="+res.direction)
    })
    wx.stopCompass() //停止监听罗盘数据
  }
})
```

4.8.5 拨打电话

微信小程序使用wx.makePhoneCall(OBJECT)来拨打电话，具体参数如表4.41所示。

表4.41 wx. makePhoneCall参数说明

属性	类型	是否必填	说明
phoneNumber	String	是	需要拨打的电话号码
success	Function	是	接口调用成功的回调函数
fail	Function	否	接口调用失败的回调函数
complete	Function	否	接口调用结束的回调函数（调用成功、失败都会执行）

示例代码如下所示。

```
wx.makePhoneCall({
  phoneNumber: '13811112222'
})
```

4.8.6 扫码

微信小程序使用wx.scanCode(OBJECT)来调出客户端扫码界面，扫码成功后返回对应的结果，具体参数如表4.42所示。

表4.42 wx. scanCode参数说明

属性	类型	是否必填	说明
success	Function	是	接口调用成功的回调函数，返回内容详见返回参数说明
fail	Function	否	接口调用失败的回调函数
complete	Function	否	接口调用结束的回调函数（调用成功、失败都会执行）

success返回参数说明如表4.43所示。

表4.43 success参数说明

参数	说明
result	码的内容

示例代码如下所示。

```
wx.scanCode({
  success: (res) => {
    console.log(res)
  }
})
```

4.8.7 剪贴板

微信小程序提供剪贴板的功能，可以使用wx.setClipboardData(OBJECT)设置剪贴板的内容，同时可以使用wx.getClipboardData(OBJECT)来获取剪贴板的内容。wx.setClipboardData(OBJECT)设置剪贴板的内容具体参数如表4.44所示。

表4.44 wx. setClipboardData参数说明

属性	类型	是否必填	说明
success	Function	是	接口调用成功的回调函数，返回内容详见返回参数说明
fail	Function	否	接口调用失败的回调函数
complete	Function	否	接口调用结束的回调函数（调用成功、失败都会执行）
data	String	是	需要设置的内容

示例代码如下所示。

```
Page({
  onLoad: function () {
    wx.setClipboardData({
      data: '我是剪贴板内容',
      complete: function (res) {
        wx.getClipboardData({
          success: function (res) {
            console.log(res.data)
          }
        })
      }
    })
  }
})
```

4.8.8 蓝牙

蓝牙功能是经常会用到的功能，在和第三方设备对接的时候，需要通过蓝牙进行数据传输，这时候蓝牙功能就有了用武之地，微信小程序也提供了蓝牙功能的API以供使用。

1. 初始化蓝牙设备和监听蓝牙状态

wx.openBluetoothAdapter(OBJECT) 用来初始化小程序蓝牙功能，生效周期为从调用wx.openBluetoothAdapter至调用wx.closeBluetoothAdapter或小程序被关闭为止。在小程序蓝牙适配器模块生效期间，开发者可以正常调用下面的小程序API，并会收到蓝牙模块相关的on回调。

wx.closeBluetoothAdapter(OBJECT) 关闭蓝牙模块，使其进入未初始化状态。调用该方法将断开所有已建立的链接并释放系统资源。建议在使用小程序蓝牙流程后调用，与wx.openBluetoothAdapter成对调用。

wx.onBluetoothAdapterStateChange(CALLBACK)用来监听蓝牙状态，返回值available为true时代表蓝牙适配器可用，discovering为true时为蓝牙适配器处于搜索状态。

wx.getBluetoothAdapterState(OBJECT)用来获取蓝牙状态，返回值discovering为true时为正在搜索设备，available为true时代表蓝牙适配器可用，errMsg（成功：ok。错误：详细信息）。

示例代码如下所示。

```
Page({
  onLoad: function () {
    wx.openBluetoothAdapter({   //初始化蓝牙功能
      success: function (res) {
        console.log(res)
      }
    })
    wx.closeBluetoothAdapter({   //关闭蓝牙功能
      success: function (res) {
        console.log(res)
      }
    })
    wx.onBluetoothAdapterStateChange(function (res) {   //监听蓝牙状态
      console.log('adapterState changed, now is', res)
    })
    wx.getBluetoothAdapterState({   //获取蓝牙状态
      complete: function(res) {
        console.log(res)
      }
    })
  }
})
```

2．蓝牙设备

（1）wx.startBluetoothDevicesDiscovery(OBJECT) 开始搜寻附近的蓝牙外围设备。注意，该操作比较耗费系统资源，请在搜索并连接到设备后调用 wx.stopBluetoothDevicesDiscovery(OBJECT) 停止搜寻附近的蓝牙外围设备。wx.startBluetoothDevicesDiscovery(OBJECT)具体参数如表4.45所示。

表4.45　wx. startBluetoothDevicesDiscovery参数说明

属性	类型	是否必填	说明
success	Function	是	接口调用成功的回调函数，返回内容详见返回参数说明
fail	Function	否	接口调用失败的回调函数
complete	Function	否	接口调用结束的回调函数（调用成功、失败都会执行）
services	Array	否	蓝牙设备主 service 的 uuid 列表
allowDuplicatesKey	Boolean	否	是否允许重复上报同一设备，如果允许重复上报，则onDeviceFound方法会多次上报同一设备，但是 RSSI 值会有不同
interval	Number	否	上报设备的间隔，默认为0，即找到新设备立即上报，否则根据传入的间隔上报

示例代码如下所示。

```
Page({
  onLoad: function () {
```

```
wx.startBluetoothDevicesDiscovery({
    services: ['FEE7'],
    success: function (res) {
      console.log(res)
    }
  })
  wx.stopBluetoothDevicesDiscovery({
    success: function (res) {
      console.log(res)
    }
  })
 }
})
```

（2）wx.getBluetoothDevices(OBJECT) 获取在小程序蓝牙模块生效期间所有已发现的蓝牙设备，包括已经和本机处于连接状态的设备。wx.getConnectedBluetoothDevices(OBJECT) 根据 uuid 获取处于已连接状态的设备。wx.onBluetoothDeviceFound(CALLBACK) 监听寻找到新设备的事件。

示例代码如下所示。

```
Page({
  onLoad: function () {
    function ab2hex(buffer) {
      var hexArr = Array.prototype.map.call(
        new Uint8Array(buffer),
        function (bit) {
          return ('00' + bit.toString(16)).slice(-2)
        }
      )
      return hexArr.join('');
    }
    wx.getBluetoothDevices({
      success: function (res) {
        console.log(res)
        if (res.devices[0]) {
          console.log(ab2hex(res.devices[0].advertisData))
        }
      }
    })
    wx.getConnectedBluetoothDevices({
      success: function (res) {
        console.log(res)
      }
    })
    wx.onBluetoothDeviceFound(function (devices) {
      console.log('new device list has founded')
      console.dir(devices)
      console.log(ab2hex(devices[0].advertisData))
    })
  }
})
```

（3）低功耗蓝牙设备

wx.createBLEConnection(OBJECT)用来连接低功耗蓝牙设备；wx.closeBLEConnection(OBJECT)

断开与低功耗蓝牙设备的连接；wx.onBLEConnectionStateChange(CALLBACK) 监听低功耗蓝牙连接状态的改变事件，包括开发者主动连接或断开连接、设备丢失、连接异常断开等；wx.notifyBLECharacteristicValueChange(OBJECT) 启用低功耗蓝牙设备特征值变化时的 notify 功能，订阅特征值，设备的特征值必须支持notify或者indicate才可以成功调用，具体参照 characteristic 的 properties 属性；wx.onBLECharacteristicValueChange(CALLBACK) 监听低功耗蓝牙设备的特征值变化，必须先启用notify接口才能接收到设备推送的notification；wx.readBLECharacteristicValue(OBJECT) 读取低功耗蓝牙设备的特征值的二进制数据值；wx.writeBLECharacteristicValue(OBJECT) 向低功耗蓝牙设备特征值中写入二进制数据。

示例代码如下所示。

```javascript
Page({
  onLoad: function () {
    wx.createBLEConnection({ //连接低功耗蓝牙设备
      // 这里的 deviceId 需要已经通过 createBLEConnection 与对应设备建立连接
      deviceId: 'deviceId',
      success: function (res) {
        console.log(res)
      }
    })
    wx.closeBLEConnection({ //断开与低功耗蓝牙设备的连接
      deviceId: 'deviceId',
      success: function (res) {
        console.log(res)
      }
    })
    wx.notifyBLECharacteristicValueChange({ //启用低功耗蓝牙设备特征值变化
      state: true, // 启用 notify 功能
      // 这里的 deviceId 需要已经通过 createBLEConnection 与对应设备建立连接
      deviceId: deviceId,
      // 这里的 serviceId 需要在上面的 getBLEDeviceServices 接口中获取
      serviceId: serviceId,
      // 这里的 characteristicId 需要在上面的 getBLEDeviceCharacteristics 接口中获取
      characteristicId: characteristicId,
      success: function (res) {
        console.log('notifyBLECharacteristicValueChange success', res.errMsg)
      }
    })
    wx.onBLEConnectionStateChange(function (res) { //监听低功耗蓝牙连接状态的改变事件
      // 该方法回调中可以用于处理连接意外断开等异常情况
      console.log(`device ${res.deviceId} state has changed, connected: ${res.connected}`)
    })
    function ab2hex(buffer) {
      var hexArr = Array.prototype.map.call(
        new Uint8Array(buffer),
        function (bit) {
          return ('00' + bit.toString(16)).slice(-2)
        }
      )
      return hexArr.join('');
    }
```

```
    wx.onBLECharacteristicValueChange(function (res) { //监听低功耗蓝牙设备的特征值变化
      console.log('characteristic ${res.characteristicId} has changed, now is ${res.value}')
      console.log(ab2hext(res.value))
    })
  }
})
```

（4）蓝牙设备服务

wx.getBLEDeviceServices(OBJECT) 用来获取蓝牙设备所有 service（服务）；wx.getBLEDeviceCharacteristics(OBJECT) 用来获取蓝牙设备某个服务中的所有 characteristic（特征值）。

示例代码如下所示。

```
Page({
  onLoad: function () {
    wx.getBLEDeviceServices({
      // 这里的 deviceId 需要已经通过 createBLEConnection 与对应设备建立连接
      deviceId: deviceId,
      success: function (res) {
        console.log('device services:', res.services)
      }
    })
    wx.getBLEDeviceCharacteristics({
      // 这里的 deviceId 需要已经通过 createBLEConnection 与对应设备建立连接
      deviceId: deviceId,
      // 这里的 serviceId 需要在上面的 getBLEDeviceServices 接口中获取
      serviceId: serviceId,
      success: function (res) {
        console.log('device getBLEDeviceCharacteristics:', res.characteristics)
      }
    })
  }
})
```

4.8.9 屏幕亮度

使用wx.setScreenBrightness(OBJECT)来设置屏幕亮度，它有一个参数值value，其范围是0~1，0代表最暗，1代表最亮；使用wx.getScreenBrightness(OBJECT)获取屏幕的亮度；wx.setKeepScreenOn(OBJECT)用来设置是否保持常亮状态，仅在当前小程序生效，离开小程序后设置失效，它有一个是否保持屏幕长亮keepScreenOn的参数设置。

4.8.10 用户截屏事件

使用wx.onUserCaptureScreen(CALLBACK)，监听用户主动截屏事件，用户使用系统截屏按键截屏时触发此事件。

示例代码如下所示。

```
wx.onUserCaptureScreen(function(res) {
    console.log('用户截屏了')
})
```

4.8.11 振动

使用wx.vibrateLong(OBJECT) 使手机发生较长时间的振动（400ms）；使用wx.vibrateShort(OBJECT)使手机发生较短时间的振动（15ms）。

4.8.12 手机联系人

调用wx.addPhoneContact(OBJECT)后，用户可以选择将该表单以"新增联系人"或"添加到已有联系人"的方式，写入手机系统通信录，完成手机通信录联系人和联系方式的增加。具体参数如表4.46所示。

表4.46 wx. addPhoneContact (OBJECT)参数说明

属性	类型	是否必填	说明
photoFilePath	String	否	头像本地文件路径
nickName	String	否	昵称
lastName	String	否	姓氏
middleName	String	否	中间名
firstName	String	是	名字
Remark	String	否	备注
mobilePhoneNumber	String	否	手机号
weChatNumber	String	否	微信号
addressCountry	String	否	联系地址国家
addressState	String	否	联系地址省份
addressCity	String	否	联系地址城市
addressStreet	String	否	联系地址街道
addressPostalCode	String	否	联系地址邮政编码
organization	String	否	公司
Title	String	否	职位
workFaxNumber	String	否	工作传真
workPhoneNumber	String	否	工作电话
hostNumber	String	否	公司电话
Email	String	否	电子邮件
url	String	否	网站
workAddressCountry	String	否	工作地址国家
workAddressState	String	否	工作地址省份
workAddressCity	String	否	工作地址城市
workAddressStreet	String	否	工作地址街道
workAddressPostalCode	String	否	工作地址邮政编码
homeFaxNumber	String	否	住宅传真
homePhoneNumber	String	否	住宅电话
homeAddressCountry	String	否	住宅地址国家
homeAddressState	String	否	住宅地址省份
homeAddressCity	String	否	住宅地址城市
homeAddressStreet	String	否	住宅地址街道
homeAddressPostalCode	String	否	住宅地址邮政编码
success	Function	否	接口调用成功的回调函数，返回内容详见返回参数说明
fail	Function	否	接口调用失败的回调函数
complete	Function	否	接口调用结束的回调函数（调用成功、失败都会执行）

回调结果如表4.47所示。

表4.47 回调结果

回调类型	errMsg	说明
success	Ok	添加成功
fail	fail cancel	用户取消操作
fail	fail ${detail}	调用失败，detail 加上详细信息

示例代码如下所示。

```
Page({
  onLoad:function(){
    wx.addPhoneContact({
      firstName:'名字',
      nickName:'昵称',
      lastName:'姓氏',
      mobilePhoneNumber:'手机号'
    })
  }
})
```

4.9 交互反馈API

交互反馈API

微信小程序在用户操作过程中，提供了4种交互反馈API：wx.showToast(OBJECT) 显示消息提示框、wx.hideToast()隐藏消息提示框、wx.showModal(OBJECT) 显示模态弹窗、wx.showActionSheet(OBJECT) 显示操作菜单。

4.9.1 消息提示框

消息提示框是一种经常用来提交成功或者加载中的友好提示方式，如图4.24、图4.25所示。

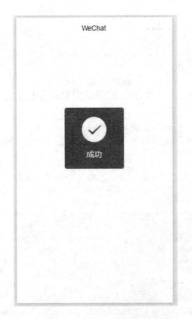

图4.24 success类型

图4.25 loading类型

消息提示框可以设置提示框的内容、类型、时间以及相应的事件,如果想显示消息提示框,可以使用wx.showToast(OBJECT)的API,wx.showToast(OBJECT)的具体参数如表4.48所示。

表4.48 wx.showToast参数说明

属性	类型	是否必填	说明
title	String	是	提示的内容
icon	String	否	图标,只支持"success""loading"
duration	Number	否	提示的延迟时间,单位为毫秒,默认值为1500,最大为10000
mask	Boolean	否	是否显示透明蒙层,防止触摸穿透,默认为false
success	Function	否	接口调用成功的回调函数
fail	Function	否	接口调用失败的回调函数
complete	Function	否	接口调用结束的回调函数(调用成功、失败都会执行)

示例代码如下所示。

```
Page({
  onLoad: function () {
    wx.showToast({
      title: '成功',
      icon: 'success',
      duration: 2000
    })
  }
})

Page({
  onLoad: function () {
    wx.showToast({
      title: '加载中',
      icon: 'loading',
      duration: 2000
    })
  }
})
```

如果想手动地隐藏信息提示框,可以使用wx.hideToast()这个接口来隐藏消息提示框,示例代码如下所示。

```
Page({
  onLoad: function () {
    wx.showToast({
      title: '加载中',
      icon: 'loading',
      duration: 10000
    })

    setTimeout(function () {
      wx.hideToast()
    }, 2000)
  }
})
```

4.9.2 模态弹窗

模态弹窗用来对整个界面进行覆盖，防止用户操作界面中的其他内容，如图4.26所示。

图4.26 模态弹窗

使用wx.showModal(OBJECT)显示模态弹窗，可以设置提示的标题、提示的内容、取消按钮和样式、确定按钮和样式以及一些绑定的事件，具体参数说明如表4.49所示。

表4.49 wx. ShowModal参数说明

属性	类型	是否必填	说明
title	String	是	提示的标题
content	String	是	提示的内容
showCancel	Boolean	否	是否显示取消按钮，默认为 true
cancelText	String	否	取消按钮的文字，默认为"取消"，最多为4个字符
cancelColor	HexColor	否	取消按钮的文字颜色，默认为"#000000"
confirmText	String	否	确定按钮的文字，默认为"确定"，最多为4个字符
confirmColor	HexColor	否	确定按钮的文字颜色，默认为"#3CC51F"
success	Function	否	接口调用成功的回调函数，返回res.confirm为true时，表示用户点击确定按钮
fail	Function	否	接口调用失败的回调函数
complete	Function	否	接口调用结束的回调函数（调用成功、失败都会执行）

示例代码如下所示。

```
Page({
  onLoad: function () {
    wx.showModal({
      title: '提示',
      content: '这是一个模态弹窗',
      success: function (res) {
```

```
            if (res.confirm) {
                console.log('用户点击确定')
            }
        }
    })
   }
})
```

4.9.3 操作菜单

在App软件里,经常会看到从底部弹出很多选项供我们选择,同时也可以取消选择,如图4.27所示。

图4.27 操作菜单

在微信小程序里,实现这样的效果需要使用wx.showActionSheet(OBJECT)显示操作菜单这个API接口,具体参数如表4.50所示。

表4.50 wx.showActionSheet参数说明

属性	类型	是否必填	说明
itemList	String Array	是	按钮的文字数组,数组长度最大为6个
itemColor	HexColor	否	按钮的文字颜色,默认为"#000000"
success	Function	否	接口调用成功的回调函数,详见返回参数说明
fail	Function	否	接口调用失败的回调函数
complete	Function	否	接口调用结束的回调函数(调用成功、失败都会执行)

示例代码如下所示。

```
Page({
  onLoad: function () {
    wx.showActionSheet({
      itemList: ['语文', '数学', '英语', '化学', '物理', '生物',],
      success: function (res) {
        if (!res.cancel) {
```

```
            console.log(res.tapIndex)
        }
      }
    })
  }
})
```

4.10 登录API

登录API

微信小程序的登录是必不可少的环节，可以简单理解为以下几个步骤：

（1）在微信小程序里使用wx.login方法获取登录凭证code值；

（2）拿到code值后再加上AppID、secret（公众开发平台AppID下面显示的内容）、grant_type授权类型这3个参数发送到自己后台开发的服务器上，在后台服务器上去请求https://api.weixin.qq.com/sns/jscode2session这个路径，同时传递这3个参数，来获取唯一标识（openid）和会话密钥（session_key）；

（3）拿到唯一标识（openid）和会话密钥（session_key）后在自己后台开发的服务器上生成自己的sessionId；

（4）微信小程序可以将服务器生成的sessionId信息保存到本地缓存信息Storage里；

（5）后续用户进入微信小程序，先从Storage获得sessionId，将这个sessionId传输到服务器上进行查询来维护登录状态。

它的时序图如图4.28所示。

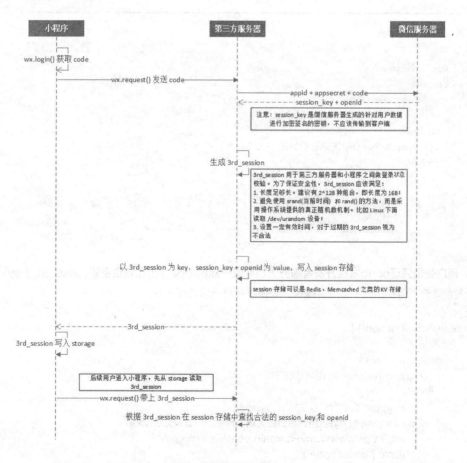

图4.28　登录时序图

1. wx.login(OBJECT)获取登录凭证code

微信小程序使用wx.login接口来获取登录凭证（code），进而获取用户登录状态信息，包括用户的唯一标识（openid）及本次登录的会话密钥（session_key）。用户数据的加解密通信需要依赖会话密钥完成。具体参数如表4.51所示。

表4.51　wx. login参数说明

属性	类型	是否必填	说明
success	Function	否	接口调用成功的回调函数
fail	Function	否	接口调用失败的回调函数
complete	Function	否	接口调用结束的回调函数（调用成功、失败都会执行）

success返回参数说明如表4.52所示。

表4.52　success参数说明

参数	类型	说明
errMsg	String	接口调用结果
code	String	用户允许登录后，回调内容会带上code（有效期为5分钟），开发者需要将code发送到开发者服务器后台，使用code 获取 session_key api，将 code 获成 openid 和 session_key

示例代码如下所示。

```
App({
  onLaunch: function() {
    wx.login({
      success:function(res){
        var code = res.code; //用户登录凭证
        if(code){
          console.log('获取用户登录凭证:'+code);
        }else{
          console.log('获取用户登录凭证失败');
        }
      }
    })
  }
})
```

2. 用户登录凭证code发往开发者服务器换取唯一标识(openid)和会话密钥(session_key)

开发者服务器需要提供一个后台接口，来接收用户登录凭证code。

```
App({
  onLaunch: function() {
    wx.login({
      success:function(res){
        var code = res.code; //用户登录凭证
        if(code){
          console.log('获取用户登录凭证:'+code);
          wx.request({ //请求自己后台服务器，传输用户登录凭证code
            url: 'https://www.my-domain.com/wx/onlogin',
            data: { code: code }
          })
```

```
            }else{
                console.log('获取用户登录凭证失败');
            }
          }
        })
      }
    })
```

3. 开发者服务器获取唯一标识(openid)和会话密钥(session_key)

开发者服务器接收到用户登录凭证code值后，加上小程序唯一标识appid、小程序secret、小程序的授权类型grant_type 3个参数去请求微信服务器接口https://api.weixin.qq.com/sns/jscode2session，来获取会话密钥 session_key 和唯一标识 openid。其中 session_key 是对用户数据进行加密签名的密钥。为了自身应用安全，session_key 不要在网络上传输。

接口地址为https://api.weixin.qq.com/sns/jscode2session?Appid=APPID&secret=SECRET&js_code=JSCODE&grant_type=authorization_code。

参数说明如表4.53所示。

表4.53　接口参数说明

属性	是否必填	说明
appid	是	小程序唯一标识
secret	是	小程序的 App secret
js_code	是	登录时获取的 code
grant_type	是	填写为 authorization_code

返回参数说明如表4.54所示。

表4.54　返回参数说明

参数	说明
openid	用户唯一标识
session_key	会话密钥

后台服务器有多种语言实现方式来请求微信服务器代码实现，可以是Java语言、PHP语言、Node.js语言等不同的实现方式。

下面代码使用Java语言利用Http请求微信服务器来获取用户唯一标识（openid）和会话密钥（session_key）。

```java
@Controller
@RequestMapping("/login")
public class CopyOfWxLoginAction {

    //小程序appid
    public static final String appid = "";
    //微信服务器会话验证
    public static final String access_token_url = "https://api.weixin.qq.com/sns/jscode2session";
    //授权类型
    public static final String grant_type = "authorization_code";
    //小程序密钥
    public static final String app_secret = "";
```

```java
@RequestMapping("/wxlogin")
@ResponseBody
public static Map<String, Object> login(String code) throws Exception {
    if (code == null || code.equals("")) {
        throw new Exception("小程序登录凭证code值不能为空");
    }
    //返回结果
    Map<String, Object> ret = new HashMap<String, Object>();

    Map<String,String> param = new HashMap<String,String>();
    param.put("grant_type", grant_type);
    param.put("appid", appid);
    param.put("secret", app_secret);
    param.put("js_code", code);

    //调用获取access_token接口
    String result = httpPost(access_token_url,param);
    //返回正确值：{"session_key":"mg0QgRK+BYcg4llf0sRaxQ==","expires_in":7200,"openid": "oXSIY0cGF7-8JaJIdPgllCB-BGGo"}
    //返回错误值：{"errcode":40163,"errmsg":"code been used, hints: [ req_id: KV5qfa0875th31 ]"}
    //根据请求结果判定，是否验证成功
    JSONObject obj = JSONObject.fromObject(result);
    if(obj != null){
     Object errCode = obj.get("errcode");
     if(errCode != null){
         throw new Exception("errCode:"+errCode);
     }else{
         Object session_key = obj.get("session_key"); //唯一标识(openid)
         Object openid = obj.get("openid"); //会话密钥(session_key)
         ret.put("session_key", session_key);
         ret.put("openid", openid);
     }
    }

    return ret;
}

public static String httpGet(String url) {
    DefaultHttpClient httpclient = new DefaultHttpClient();
    String body = null;
    HttpGet get = new HttpGet(url);
    body = invoke(httpclient, get);

    httpclient.getConnectionManager().shutdown();

    return body;
}

public static String httpPost(String url, Map<String, String> params) {
    DefaultHttpClient httpclient = new DefaultHttpClient();
    String body = null;
```

```java
        HttpPost post = postForm(url, params);
        body = invoke(httpclient, post);

        httpclient.getConnectionManager().shutdown();

        return body;
    }

    private static HttpPost postForm(String url, Map<String, String> params){

        HttpPost httpost = new HttpPost(url);
        List<NameValuePair> nvps = new ArrayList <NameValuePair>();

        Set<String> keySet = params.keySet();
        for(String key : keySet) {
            nvps.add(new BasicNameValuePair(key, params.get(key)));
        }

        try {
            httpost.setEntity(new UrlEncodedFormEntity(nvps, HTTP.UTF_8));
        } catch (UnsupportedEncodingException e) {
            e.printStackTrace();
        }

        return httpost;
    }

    private static String invoke(DefaultHttpClient httpclient,
            HttpUriRequest httpost) {

        HttpResponse response = sendRequest(httpclient, httpost);
        String body = paseResponse(response);

        return body;
    }

    private static HttpResponse sendRequest(DefaultHttpClient httpclient,
            HttpUriRequest httpost) {
        HttpResponse response = null;

        try {
            response = httpclient.execute(httpost);
        } catch (ClientProtocolException e) {
            e.printStackTrace();
        } catch (IOException e) {
            e.printStackTrace();
        }
        return response;
    }

    private static String paseResponse(HttpResponse response) {
        HttpEntity entity = response.getEntity();
```

```
                String charset = EntityUtils.getContentCharSet(entity);
                charset = Utils.isEmpty(charset) ? "utf-8" : charset;
                String body = null;
                try {
                    body = EntityUtils.toString(entity, charset);
                } catch (ParseException e) {
                    e.printStackTrace();
                } catch (IOException e) {
                    e.printStackTrace();
                }

                return body;
            }
        }
```

4. 开发者服务器生成自己的sessionId

开发者服务器获取到唯一标识（openid）和会话密钥（session_key）后，需要生成自己的sessionId，规则可以由自己确定，可以采用拼接成字符串或者拼接成字符串然后再MD5加密等多种方式。生成的sessionId需要在开发者服务器中保存起来，小程序在校验登录或者有一些操作是在登录后才可以进行的，都需要到开发者服务器来验证。sessionId保存到缓存Memcached、Redis、内存里都是可以的。

5. 小程序客户端保存sessionId

小程序客户端是没有类似于浏览器客户端的cookie或者session的机制，但是可以利用小程序的storage缓存机制来保存sessionId，在后续需要登录状态请求时可以使用。在之后调用那些需要登录后才有权限访问的后台服务时，可以将保存在storage中的sessionId取出并携带在请求中，传递到后台服务，后台代码获取到该sessionId后，从缓存Redis或者内存中查找是否有该sessionId存在，如果有的话，即确认该session是有效的，可继续后续的代码执行，否则要进行错误处理。

6. wx.checkSession(OBJECT) 检查登录状态是否过期

微信小程序可以使用wx.checkSession(OBJECT)检查登录状态是否过期，如果过期就需要重新登录，具体参数如表4.55所示。

表4.55　wx.checkSession参数说明

属性	类型	是否必填	说明
success	Function	否	接口调用成功的回调函数，登录状态未过期
fail	Function	否	接口调用失败的回调函数，登录状态已过期
complete	Function	否	接口调用结束的回调函数（调用成功、失败都会执行）

示例代码如下所示。

```
wx.checkSession({
  success: function(){
    //登录状态未过期
  },
  fail: function(){
    //登录状态已过期
    wx.login()
  }
})
```

7. wx.getUserInfo(OBJECT) 获取用户信息

微信小程序使用wx.getUserInfo(OBJECT)来获取用户信息，在获取用户信息之前，需要调用wx.login接口，用户只有在登录状态才能获取到用户的相关信息。具体参数如表4.56所示。

表4.56 wx.getUserInfo参数说明

属性	类型	是否必填	说明
success	Function	否	接口调用成功的回调函数
fail	Function	否	接口调用失败的回调函数
complete	Function	否	接口调用结束的回调函数（调用成功、失败都会执行）

success返回参数如表4.57所示。

表4.57 success返回参数说明

属性	类型	说明
userInfo	OBJECT	用户信息对象，不包含 openid 等敏感信息
rawData	String	不包括敏感信息的原始数据字符串，用于计算签名
signature	String	使用 sha1(rawData + sessionkey) 得到字符串，用于校验用户信息
encryptedData	String	包括敏感数据在内的完整用户信息的加密数据，详细见加密数据解密算法
iv	String	加密算法的初始向量，详细见加密数据解密算法

示例代码如下所示。

```
Page({
  onLoad: function () {
    wx.getUserInfo({
      success: function (res) {
        console.log(res);
        var userInfo = res.userInfo
        var nickName = userInfo.nickName
        var avatarUrl = userInfo.avatarUrl
        var gender = userInfo.gender //性别 0：未知。1：男。2：女
        var province = userInfo.province
        var city = userInfo.city
        var country = userInfo.country
      }
    })
  }
})
```

4.11 微信支付API

微信支付API

微信支付有很多种方式，有刷卡支付、扫码支付、公众号支付、App支付和小程序支付，我们下面讲解的都是微信小程序的支付方式。

4.11.1 微信小程序支付介绍

微信支付主要经过小程序内调用登录接口、商户server调用支付统一下单、商户server调用再次签名、商户server接收支付通知、商户server查询支付结果这几个过程。

微信小程序支付实现步骤：

（1）微信小程序调用wx.login方法，获取用户登录凭证code；

（2）微信小程序将用户登录凭证code传输给自己的开发者后台服务器；

（3）开发者后台服务器根据用户登录凭证code向微信服务器请求获取唯一标识（openid）；

（4）开发者后台服务器获取到唯一标识（openid）后，调用统一下单支付接口，来获取预支付交易会话标识（prepay_id）；

（5）开发者后台服务器调用签名，并返回支付需要使用的参数；

（6）微信小程序调用wx.requestPayment方法发起微信支付；

（7）开发者后台服务器接收微信服务器的通知并处理微信服务器返回的结果。

小程序支付交互过程如图4.29所示。

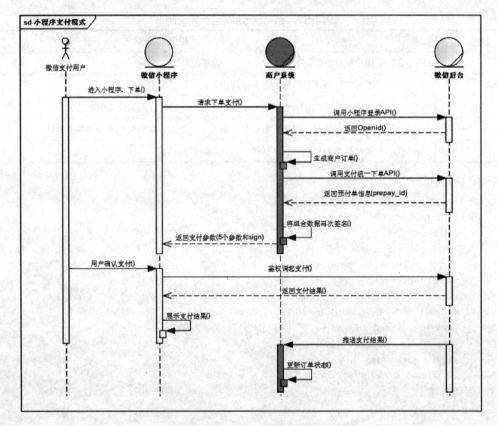

图4.29 小程序支付交互时序图

说明：

（1）商户server调用支付统一下单，API参见公共API：https://pay.weixin.qq.com/wiki/doc/api/wxa/wxa_api.php?chapter=9_1；

（2）商户server调用再次签名，API参见公共API：https://pay.weixin.qq.com/wiki/doc/api/wxa/wxa_api.php?chapter=7_7；

（3）商户server接收支付通知，API参见公共API：https://pay.weixin.qq.com/wiki/doc/api/wxa/wxa_api.php?chapter=9_7；

（4）商户server查询支付结果，API参见公共API：https://pay.weixin.qq.com/wiki/doc/api/wxa/

wxa_api.php?chapter=9_2。

微信小程序提供了微信支付接口，可以使用wx.requestPayment(OBJECT)来进行微信支付，具体参数说明如表4.58所示。

表4.58　wx.requestPayment小程序支付

属性	类型	是否必填	说明
timeStamp	String	是	时间戳从1970年1月1日00:00:00至今的秒数,即当前的时间
nonceStr	String	是	随机字符串,长度为32个字符以下
package	String	是	统一下单接口返回的 prepay_id 参数值,提交格式如：prepay_id=*
signType	String	是	签名算法,暂支持 MD5
paySign	String	是	签名,具体签名方案参见微信公众号支付帮助文档
success	Function	否	接口调用成功的回调函数
fail	Function	否	接口调用失败的回调函数
complete	Function	否	接口调用结束的回调函数（调用成功、失败都会执行）

示例代码如下所示：

```
wx.requestPayment({
  'timeStamp': '',
  'nonceStr': '',
  'package': '',
  'signType': 'MD5',
  'paySign': '',
  'success':function(res){
  },
  'fail':function(res){
  }
})
```

4.11.2　微信小程序支付实战

微信小程序支付实现步骤：

（1）微信小程序调用wx.login方法，获取用户登录凭证code；

（2）微信小程序将用户登录凭证code传输给自己的开发者后台服务器；

（3）开发者后台服务器根据用户登录凭证code向微信服务器请求获取唯一标识（openid）；

（4）开发者后台服务器获取到唯一标识（openid）后，开发者后台服务器调用统一下单支付接口，来获取预支付交易会话标识（prepay_id）；

（5）开发者后台服务器调用签名，并返回支付需要使用的参数；

（6）微信小程序调用wx.requestPayment方法发起微信支付；

（7）开发者后台服务器接收微信服务器的通知并处理微信服务器返回的结果。

1．获取用户登录凭证并传输给开发者后台服务器

```
Page({
  onLoad:function(){
    wx.login({
      success: function(res) {
        var code = res.code;//用户登录凭证code
        if (code) {
```

```
            //发起网络请求,将用户登录凭证传输给开发者后台服务器
            wx.request({
              url: 'https://test.com/wxapp/WxLoginServlet',
              data: {
                code: code
              }
            })
          } else {
            console.log('登录失败！' + res.errMsg)
          }
        }
      });
    }
  })
```

2. 开发者后台服务器接收用户登录凭证并向微信服务器获取唯一标识openid和会话秘钥session_key，将openid回传给微信小程序客户端

（1）定义一个微信支付参数配置类WxPayConfig.java。

```
package com.xiaogang.wxpay;

/**
 * 微信小程序支付配置
 * @author xiaogang
 *
 */
public class WxPayConfig {

    //小程序appid
    public static final String appid = "";
    //微信支付的商户id
    public static final String mch_id = "";
    //微信支付的商户密钥
    public static final String key = "";
    //支付成功后的服务器回调url
    public static final String notify_url = "https://xx/xx/weixin/api/wxNotify";
    //签名方式，固定值
    public static final String SIGNTYPE = "MD5";
    //交易类型，小程序支付的固定值为JSAPI
    public static final String TRADETYPE = "JSAPI";
    //微信统一下单接口地址
    public static final String pay_url = "https://api.mch.weixin.qq.com/pay/unifiedorder";
    //微信服务器会话验证
    public static final String access_token_url = "https://api.weixin.qq.com/sns/jscode2session";
    //授权类型
    public static final String grant_type = "authorization_code";
    //小程序密钥
    public static final String app_secret = "";
}
```

（2）创建一个HttpClient工具类HttpUtil.java，可以通过Http向微信服务器发起请求。

```
package com.xiaogang.app.util;
```

```java
import java.io.IOException;
import java.io.UnsupportedEncodingException;
import java.util.ArrayList;
import java.util.List;
import java.util.Map;
import java.util.Set;

import org.apache.http.HttpEntity;
import org.apache.http.HttpResponse;
import org.apache.http.NameValuePair;
import org.apache.http.ParseException;
import org.apache.http.client.ClientProtocolException;
import org.apache.http.client.ResponseHandler;
import org.apache.http.client.entity.UrlEncodedFormEntity;
import org.apache.http.client.methods.HttpGet;
import org.apache.http.client.methods.HttpPost;
import org.apache.http.client.methods.HttpUriRequest;
import org.apache.http.impl.client.CloseableHttpClient;
import org.apache.http.impl.client.DefaultHttpClient;
import org.apache.http.impl.client.HttpClients;
import org.apache.http.message.BasicNameValuePair;
import org.apache.http.protocol.HTTP;
import org.apache.http.protocol.HttpContext;
import org.apache.http.util.EntityUtils;

/**
 * @ClassName: HttpUtil
 * @Description: 提供HttpClient调用方法，支持get和post提交
 * @author xiaogang
 * @date 2017年3月3日 下午1:24:50
 */
public class HttpUtil {

    public static String post(String url, Map<String, String> params) {
        DefaultHttpClient httpclient = new DefaultHttpClient();
        String body = null;

        HttpPost post = postForm(url, params);
        body = invoke(httpclient, post);

        httpclient.getConnectionManager().shutdown();

        return body;
    }

    public static String post(String url, Map<String, String> params,String cookie) {
        DefaultHttpClient httpclient = new DefaultHttpClient();
        String body = null;
        HttpPost post = postForm(url, params);
        post.setHeader("cookie", cookie);
        body = invoke2(httpclient, post);
```

```java
            httpclient.getConnectionManager().shutdown();

        return body;
    }
    public static String post(String url, Map<String, String> params,String cookie,HttpContext localContext) {
        DefaultHttpClient httpclient = new DefaultHttpClient();
        HttpPost post = postForm(url, params);
        post.setHeader("cookie", cookie);
        HttpResponse response = null;

            try {
                response = httpclient.execute(post,localContext);
            } catch (ClientProtocolException e) {
                // TODO Auto-generated catch block
                e.printStackTrace();
            } catch (IOException e) {
                // TODO Auto-generated catch block
                e.printStackTrace();
            }
        String location = response.getFirstHeader("Location")
                .getValue();

        httpclient.getConnectionManager().shutdown();

        return location;
    }
    public static String get(String url) {
        DefaultHttpClient httpclient = new DefaultHttpClient();
        String body = null;

        //log.info("create httppost:" + url);
        HttpGet get = new HttpGet(url);
        body = invoke(httpclient, get);

        httpclient.getConnectionManager().shutdown();

        return body;
    }
    public static String httpGet(String url){
        CloseableHttpClient httpclient = HttpClients.createDefault();
        try {
            HttpGet httpget = new HttpGet(url);
            // Create a custom response handler
            ResponseHandler<String> responseHandler = new ResponseHandler<String>() {
                public String handleResponse(
                        final HttpResponse response) throws ClientProtocolException, IOException {
                    int status = response.getStatusLine().getStatusCode();
                    if (status >= 200 && status < 300) {
                        HttpEntity entity = response.getEntity();
                        return entity != null ? EntityUtils.toString(entity, "utf-8") : null;
                    } else {
```

```java
                    throw new ClientProtocolException("Unexpected response status: " + status);
                }
            }

        };
        return httpclient.execute(httpget, responseHandler);

    } catch (Exception e) {
        e.printStackTrace();
    }finally {
        try {
            httpclient.close();
        } catch (IOException e) {
            e.printStackTrace();
        }
    }
    return null;
}

private static String invoke(DefaultHttpClient httpclient,
        HttpUriRequest httpost) {

    HttpResponse response = sendRequest(httpclient, httpost);
    String body = paseResponse(response);

    return body;
}
private static String invoke2(DefaultHttpClient httpclient,
        HttpUriRequest httpost) {

    HttpResponse response = sendRequest(httpclient, httpost);
    if(null==response.getFirstHeader("Location")){
        return response.getStatusLine().toString();
    }else{
    String location = response.getFirstHeader("Location")
            .getValue();
    return location;
    }
}


private static String paseResponse(HttpResponse response) {
    //log.info("get response from http server..");
    HttpEntity entity = response.getEntity();

    String charset = EntityUtils.getContentCharSet(entity);
    //log.info(charset);
    charset = Utils.isEmpty(charset) ? "utf-8" : charset;
    String body = null;
    try {
        body = EntityUtils.toString(entity, charset);
        //log.info(body);
```

```java
        } catch (ParseException e) {
            e.printStackTrace();
        } catch (IOException e) {
            e.printStackTrace();
        }

        return body;
    }

    private static HttpResponse sendRequest(DefaultHttpClient httpclient,
            HttpUriRequest httpost) {
        //log.info("execute post...");
        HttpResponse response = null;

        try {
            response = httpclient.execute(httpost);
        } catch (ClientProtocolException e) {
            e.printStackTrace();
        } catch (IOException e) {
            e.printStackTrace();
        }
        return response;
    }

    private static HttpPost postForm(String url, Map<String, String> params){

        HttpPost httpost = new HttpPost(url);
        List<NameValuePair> nvps = new ArrayList <NameValuePair>();

        Set<String> keySet = params.keySet();
        for(String key : keySet) {
            nvps.add(new BasicNameValuePair(key, params.get(key)));
        }

        try {
            httpost.setEntity(new UrlEncodedFormEntity(nvps, HTTP.UTF_8));
        } catch (UnsupportedEncodingException e) {
            e.printStackTrace();
        }

        return httpost;
    }

}
```

（3）使用WxLoginServelt.java类来接收用户凭证code值，并向微信服务器发送请求来获取唯一标识openid和会话密钥session_key，将它们保存到内存中。

```java
package com.xiaogang.wxpay;

import java.io.IOException;
import java.util.HashMap;
```

```java
import java.util.Map;

import javax.servlet.ServletException;
import javax.servlet.http.HttpServlet;
import javax.servlet.http.HttpServletRequest;
import javax.servlet.http.HttpServletResponse;

import net.sf.json.JSONObject;

import com.xiaogang.app.util.HttpUtil;
import com.xiaogang.wxpay.WxLoginAction.WxPayConfig;

/**
 * 微信登录验证
 * @author kevin
 *
 */
public class WxLoginServlet extends HttpServlet {
    private static final long serialVersionUID = 1L;


    public WxLoginServlet() {
        super();
    }


    protected void doGet(HttpServletRequest request, HttpServletResponse response) throws ServletException, IOException {
        doPost(request, response);
    }

    /**
     * 根据用户登录凭证code来获取唯一标识openid和会话密钥session_key
     */
    protected void doPost(HttpServletRequest request, HttpServletResponse response) throws ServletException, IOException {
        String code = request.getParameter("code");
        if (code != null && !"".equals(code)) {
         code = "081qWBKh0ndG7y1q5sMh0eaTKh0qWBKd";
            Map<String, Object> ret = new HashMap<String, Object>();

            Map<String,String> param = new HashMap<String,String>();
            param.put("grant_type", WxPayConfig.grant_type);
            param.put("appid", WxPayConfig.appid);
            param.put("secret", WxPayConfig.app_secret);
            param.put("js_code", code);

            //调用获取access_token接口
            String result = HttpUtil.post(WxPayConfig.access_token_url,param);
            //返回正确值：{"session_key":"mg0QgRK+BYcg4lIf0sRaxQ==","expires_in":7200,"openid":"oXSIY0cGF7-8JaJIdPgllCB-BGGo"}
            //返回错误值：{"errcode":40163,"errmsg":"code been used, hints: [ req_id: KV5qfa0875th31 ]"}
            System.out.println(result);
            //根据请求结果判定，是否验证成功
```

```java
                            JSONObject obj = JSONObject.fromObject(result);
                            if(obj != null){
                                Object errCode = obj.get("errcode");
                                if(errCode == null){
                                            Object session_key = obj.get("session_key");
                                            Object openid = obj.get("openid");
                                            //会话密钥session_key放入内存中
                                            request.getSession().setAttribute("session_key", session_key);
                                            //唯一标识openid放入内存中
                                            request.getSession().setAttribute("openid", openid);
                                            PrintWriter out = response.getWriter();
                                            out.write(openid.toString());//将唯一标识回传给微信小程序客户端
                                }
                            }
                    }
            }
    }

}

package com.xiaogang.app.util;

import java.io.IOException;
import java.io.UnsupportedEncodingException;
import java.util.ArrayList;
import java.util.List;
import java.util.Map;
import java.util.Set;

import org.apache.http.HttpEntity;
import org.apache.http.HttpResponse;
import org.apache.http.NameValuePair;
import org.apache.http.ParseException;
import org.apache.http.client.ClientProtocolException;
import org.apache.http.client.ResponseHandler;
import org.apache.http.client.entity.UrlEncodedFormEntity;
import org.apache.http.client.methods.HttpGet;
import org.apache.http.client.methods.HttpPost;
import org.apache.http.client.methods.HttpUriRequest;
import org.apache.http.impl.client.CloseableHttpClient;
import org.apache.http.impl.client.DefaultHttpClient;
import org.apache.http.impl.client.HttpClients;
import org.apache.http.message.BasicNameValuePair;
import org.apache.http.protocol.HTTP;
import org.apache.http.protocol.HttpContext;
import org.apache.http.util.EntityUtils;


/**
 * @ClassName: HttpUtil
 * @Description: 提供HttpClient调用方法，支持get和post提交
 * @author xiaogang
 * @date 2017年3月3日 下午1:24:50
```

```java
*/
public class HttpUtil {

    public static String post(String url, Map<String, String> params) {
        DefaultHttpClient httpclient = new DefaultHttpClient();
        String body = null;

        HttpPost post = postForm(url, params);
        body = invoke(httpclient, post);

        httpclient.getConnectionManager().shutdown();

        return body;
    }

    public static String post(String url, Map<String, String> params, String cookie) {
        DefaultHttpClient httpclient = new DefaultHttpClient();
        String body = null;
        HttpPost post = postForm(url, params);
        post.setHeader("cookie", cookie);
        body = invoke2(httpclient, post);

        httpclient.getConnectionManager().shutdown();

        return body;
    }

    public static String post(String url, Map<String, String> params, String cookie, HttpContext localContext) {
        DefaultHttpClient httpclient = new DefaultHttpClient();
        HttpPost post = postForm(url, params);
        post.setHeader("cookie", cookie);
        HttpResponse response = null;

        try {
            response = httpclient.execute(post, localContext);
        } catch (ClientProtocolException e) {
            // TODO Auto-generated catch block
            e.printStackTrace();
        } catch (IOException e) {
            // TODO Auto-generated catch block
            e.printStackTrace();
        }
        String location = response.getFirstHeader("Location")
                .getValue();

        httpclient.getConnectionManager().shutdown();

        return location;
    }
    public static String get(String url) {
        DefaultHttpClient httpclient = new DefaultHttpClient();
        String body = null;
```

```java
        //log.info("create httppost:" + url);
        HttpGet get = new HttpGet(url);
        body = invoke(httpclient, get);

        httpclient.getConnectionManager().shutdown();

        return body;
}
    public static String httpGet(String url){
        CloseableHttpClient httpclient = HttpClients.createDefault();
        try {
            HttpGet httpget = new HttpGet(url);
            // Create a custom response handler
            ResponseHandler<String> responseHandler = new ResponseHandler<String>() {
                public String handleResponse(
                        final HttpResponse response) throws ClientProtocolException, IOException {
                    int status = response.getStatusLine().getStatusCode();
                    if (status >= 200 && status < 300) {
                        HttpEntity entity = response.getEntity();
                        return entity != null ? EntityUtils.toString(entity, "utf-8") : null;
                    } else {
                        throw new ClientProtocolException("Unexpected response status: " + status);
                    }
                }

            };
            return httpclient.execute(httpget, responseHandler);

        } catch (Exception e) {
            e.printStackTrace();
        }finally {
            try {
                httpclient.close();
            } catch (IOException e) {
                e.printStackTrace();
            }
        }
        return null;
}

private static String invoke(DefaultHttpClient httpclient,
            HttpUriRequest httpost) {

    HttpResponse response = sendRequest(httpclient, httpost);
    String body = paseResponse(response);

    return body;
}
private static String invoke2(DefaultHttpClient httpclient,
            HttpUriRequest httpost) {

    HttpResponse response = sendRequest(httpclient, httpost);
```

```java
            if(null==response.getFirstHeader("Location")){
                return response.getStatusLine().toString();
            }else{
            String location = response.getFirstHeader("Location")
                    .getValue();
            return location;
            }
        }


    private static String paseResponse(HttpResponse response) {
        //log.info("get response from http server..");
        HttpEntity entity = response.getEntity();

        String charset = EntityUtils.getContentCharSet(entity);
        //log.info(charset);
        charset = Utils.isEmpty(charset) ? "utf-8" : charset;
        String body = null;
        try {
            body = EntityUtils.toString(entity, charset);
            //log.info(body);
        } catch (ParseException e) {
            e.printStackTrace();
        } catch (IOException e) {
            e.printStackTrace();
        }

        return body;
    }

    private static HttpResponse sendRequest(DefaultHttpClient httpclient,
            HttpUriRequest httpost) {
        //log.info("execute post...");
        HttpResponse response = null;

        try {
            response = httpclient.execute(httpost);
        } catch (ClientProtocolException e) {
            e.printStackTrace();
        } catch (IOException e) {
            e.printStackTrace();
        }
        return response;
    }

    private static HttpPost postForm(String url, Map<String, String> params){

        HttpPost httpost = new HttpPost(url);
        List<NameValuePair> nvps = new ArrayList<NameValuePair>();

        Set<String> keySet = params.keySet();
        for(String key : keySet) {
```

```
                nvps.add(new BasicNameValuePair(key, params.get(key)));
            }

            try {
                httpost.setEntity(new UrlEncodedFormEntity(nvps, HTTP.UTF_8));
            } catch (UnsupportedEncodingException e) {
                e.printStackTrace();
            }

            return httpost;
        }
}
```

3. 微信小程序客户端发起支付请求到开发者后台服务器,将唯一标识openid传输给开发者后台服务器

```
        wx.request({
            url: url: 'https://test.com/wxapp/WxPayServlet',
            data: {
                openid: openid
            }
            method: 'GET',
            success: function (res) {
                console.log(res);
            }
        });
```

4. 开发者后台服务器接收唯一标识openid,调用微信支付统一下单接口

```
package com.xiaogang.wxpay;

import java.io.IOException;
import java.io.PrintWriter;
import java.util.HashMap;
import java.util.Map;

import javax.servlet.ServletException;
import javax.servlet.http.HttpServlet;
import javax.servlet.http.HttpServletRequest;
import javax.servlet.http.HttpServletResponse;

import com.alibaba.fastjson.JSON;
import com.xiaogang.app.util.PayUtil;

/**
 * 发起微信支付
 * @author xiaogang
 *
 */
public class WxPayServelt extends HttpServlet {
    private static final long serialVersionUID = 1L;


    public WxPayServelt() {
```

```java
        super();
    }

    protected void doGet(HttpServletRequest request, HttpServletResponse response) throws ServletException, IOException {
        doPost(request, response);
    }


    protected void doPost(HttpServletRequest request, HttpServletResponse response) throws ServletException, IOException {
        try{
            //生成32位长度的随机字符串
            String nonce_str = PayUtil.getRandomString(32);
            //商品名称
            String body = "商品名称";
            //会话唯一标识
            String openid = request.getParameter("openid");
            String spbill_create_ip = PayUtil.getIpAddr(request);
            //组装参数，用户生成统一下单接口的签名
            Map<String, String> packageParams = new HashMap<String, String>();
            packageParams.put("appid", WxPayConfig.appid);
            packageParams.put("mch_id", WxPayConfig.mch_id);
            packageParams.put("nonce_str", nonce_str);
            packageParams.put("body", body);
            packageParams.put("out_trade_no", "123456789");//商户订单号
            packageParams.put("total_fee", "1");//支付金额，这边需要转成字符串类型，否则后面的签名会失败
            packageParams.put("spbill_create_ip", spbill_create_ip);
            packageParams.put("notify_url", WxPayConfig.notify_url);//支付成功后的回调地址
            packageParams.put("trade_type", WxPayConfig.TRADETYPE);//支付方式
            packageParams.put("openid", openid);

            String prestr = PayUtil.createLinkString(packageParams); // 把数组所有元素，按照"参数=参数值"的模式用"&"字符拼接成字符串

            //MD5运算生成签名，用于统一下单接口
            String mysign = PayUtil.sign(prestr, WxPayConfig.key, "utf-8").toUpperCase();
            String out_trade_no = "123456";
            String total_fee = "100";
            //拼接统一下单接口使用的xml数据，要将上一步生成的签名一起拼接进去
            String xml = "<xml>" + "<appid>" + WxPayConfig.appid + "</appid>"
                    + "<body><![CDATA[" + body + "]]></body>"
                    + "<mch_id>" + WxPayConfig.mch_id + "</mch_id>"
                    + "<nonce_str>" + nonce_str + "</nonce_str>"
                    + "<notify_url>" + WxPayConfig.notify_url + "</notify_url>"
                    + "<openid>" + openid + "</openid>"
                    + "<out_trade_no>" + out_trade_no + "</out_trade_no>"
                    + "<spbill_create_ip>" + spbill_create_ip + "</spbill_create_ip>"
                    + "<total_fee>" + total_fee + "</total_fee>"
                    + "<trade_type>" + WxPayConfig.TRADETYPE + "</trade_type>"
                    + "<sign>" + mysign + "</sign>"
                    + "</xml>";
```

```java
            //调用统一下单接口,并接收返回的结果
            String result = PayUtil.httpRequest(WxPayConfig.pay_url, "POST", xml);

            //将微信服务器返回的结果解析后存储在Map对象中
            Map<String,String> map = PayUtil.parseXml2Map(result);
            //返回状态码来判断统一下单接口是否调用成功
            String return_code = map.get("return_code");
            //返回给小程序端需要的参数
            Map<String, Object> param = new HashMap<String, Object>();
            if(return_code=="SUCCESS"||return_code.equals(return_code)){
                //预付单信息
                String prepay_id = (String) map.get("prepay_id");
                param.put("nonceStr", nonce_str);
                param.put("package", "prepay_id=" + prepay_id);
                //生成签名的时候需要时间戳
                Long timeStamp = System.currentTimeMillis() / 1000;
                param.put("timeStamp", System.currentTimeMillis() / 1000 + "");
                //拼接签名需要的参数
                String stringSignTemp = "appId=" + WxPayConfig.appid + "&nonceStr=" + nonce_str +
"&package=prepay_id=" + prepay_id+ "&signType=MD5&timeStamp=" + timeStamp;
                //生成签名信息
                String paySign = PayUtil.sign(stringSignTemp, WxPayConfig.key, "utf-8").toUpperCase();
                param.put("paySign", paySign);
            }
            param.put("appid", WxPayConfig.appid);
            PrintWriter out = response.getWriter();
            String json = JSON.toJSONString(param);
            out.write(json);//返回给微信小程序
        }catch(Exception e){
            e.printStackTrace();
        }
    }
}


package com.xiaogang.app.util;

import java.io.BufferedReader;
import java.io.ByteArrayInputStream;
import java.io.InputStream;
import java.io.InputStreamReader;
import java.io.OutputStream;
import java.io.UnsupportedEncodingException;
import java.net.HttpURLConnection;
import java.net.URL;
import java.security.SignatureException;
import java.util.ArrayList;
import java.util.Collections;
import java.util.HashMap;
import java.util.Iterator;
import java.util.List;
import java.util.Map;
```

```java
import java.util.Random;

import javax.servlet.http.HttpServletRequest;

import org.apache.commons.codec.digest.DigestUtils;
import org.dom4j.Document;
import org.dom4j.DocumentHelper;
import org.dom4j.Element;

import com.caucho.xml.SAXBuilder;

public class PayUtil {
    /**
     * 签名字符串
     * @param text需要签名的字符串
     * @param key 密钥
     * @param input_charset编码格式
     * @return 签名结果
     */
    public static String sign(String text, String key, String input_charset) {
        text = text + "&key=" + key;
        return DigestUtils.md5Hex(getContentBytes(text, input_charset));
    }
    /**
     * 签名字符串
     * @param text需要签名的字符串
     * @param sign 签名结果
     * @param key密钥
     * @param input_charset 编码格式
     * @return 签名结果
     */
    public static boolean verify(String text, String sign, String key, String input_charset) {
        text = text + key;
        String mysign = DigestUtils.md5Hex(getContentBytes(text, input_charset));
        if (mysign.equals(sign)) {
            return true;
        } else {
            return false;
        }
    }
    /**
     * @param content
     * @param charset
     * @return
     * @throws SignatureException
     * @throws UnsupportedEncodingException
     */
    public static byte[] getContentBytes(String content, String charset) {
        if (charset == null || "".equals(charset)) {
            return content.getBytes();
        }
        try {
```

```java
            return content.getBytes(charset);
        } catch (UnsupportedEncodingException e) {
            throw new RuntimeException("MD5签名过程中出现错误,指定的编码集不对,您目前指定的编码集是:" + charset);
        }
    }

    private static boolean isValidChar(char ch) {
        if ((ch >= '0' && ch <= '9') || (ch >= 'A' && ch <= 'Z') || (ch >= 'a' && ch <= 'z'))
            return true;
        if ((ch >= 0x4e00 && ch <= 0x7fff) || (ch >= 0x8000 && ch <= 0x952f))
            return true;// 简体中文汉字编码
        return false;
    }
    /**
     * 除去数组中的空值和签名参数
     * @param sArray 签名参数组
     * @return 去掉空值与签名参数后的新签名参数组
     */
    public static Map<String, String> paraFilter(Map<String, String> sArray) {
        Map<String, String> result = new HashMap<String, String>();
        if (sArray == null || sArray.size() <= 0) {
            return result;
        }
        for (String key : sArray.keySet()) {
            String value = sArray.get(key);
            if (value == null || value.equals("") || key.equalsIgnoreCase("sign")
                    || key.equalsIgnoreCase("sign_type")) {
                continue;
            }
            result.put(key, value);
        }
        return result;
    }
    /**
     * 把数组所有元素排序,并按照"参数=参数值"的模式用"&"字符拼接成字符串
     * @param params 需要排序并参与字符拼接的参数组
     * @return 拼接后字符串
     */
    public static String createLinkString(Map<String, String> params) {
        List<String> keys = new ArrayList<String>(params.keySet());
        Collections.sort(keys);
        String prestr = "";
        for (int i = 0; i < keys.size(); i++) {
            String key = keys.get(i);
            String value = params.get(key);
            if (i == keys.size() - 1) {// 拼接时,不包括最后一个&字符
                prestr = prestr + key + "=" + value;
            } else {
                prestr = prestr + key + "=" + value + "&";
            }
        }
```

```java
        return prestr;
}
/**
 *
 * @param requestUrl请求地址
 * @param requestMethod请求方法
 * @param outputStr参数
 */
public static String httpRequest(String requestUrl,String requestMethod,String outputStr){
    // 创建SSLContext
    StringBuffer buffer = null;
    try{
        URL url = new URL(requestUrl);
        HttpURLConnection conn = (HttpURLConnection) url.openConnection();
        conn.setRequestMethod(requestMethod);
        conn.setDoOutput(true);
        conn.setDoInput(true);
        conn.connect();
        //往服务器端写内容
        if(null !=outputStr){
            OutputStream os=conn.getOutputStream();
            os.write(outputStr.getBytes("utf-8"));
            os.close();
        }
        // 读取服务器端返回的内容
        InputStream is = conn.getInputStream();
        InputStreamReader isr = new InputStreamReader(is, "utf-8");
        BufferedReader br = new BufferedReader(isr);
        buffer = new StringBuffer();
        String line = null;
        while ((line = br.readLine()) != null) {
            buffer.append(line);
        }
        br.close();
    }catch(Exception e){
        e.printStackTrace();
    }
    return buffer.toString();
}
public static String urlEncodeUTF8(String source){
    String result=source;
    try {
        result=java.net.URLEncoder.encode(source, "UTF-8");
    } catch (UnsupportedEncodingException e) {
        // TODO Auto-generated catch block
        e.printStackTrace();
    }
    return result;
}

public static InputStream String2Inputstream(String str) {
    return new ByteArrayInputStream(str.getBytes());
```

```java
}

//生成指定长度的随机字符串
public static String getRandomString(int length){
    //定义一个字符串（A-Z，a-z，0-9）即62位;
    String str="zxcvbnmlkjhgfdsaqwertyuiopQWERTYUIOPASDFGHJKLZXCVBNM1234567890";
    //由Random生成随机数
    Random random=new Random();
    StringBuffer sb=new StringBuffer();
    //长度为几就循环几次
    for(int i=0; i<length; ++i){
        //产生0-61的数字
        int number=random.nextInt(62);
        //将产生的数字通过length次承载到sb中
        sb.append(str.charAt(number));
    }
    //将承载的字符转换成字符串
    return sb.toString();
}


//获取真实的ip地址
public static String getIpAddr(HttpServletRequest request) {
    String ip = request.getHeader("X-Forwarded-For");
    if(ip != null && !"".equals(ip)   && !"unKnown".equalsIgnoreCase(ip)){
        int index = ip.indexOf(",");
        if(index != -1){
            return ip.substring(0,index);
        }else{
            return ip;
        }
    }
    ip = request.getHeader("X-Real-IP");
    if(ip != null && !"".equals(ip) && !"unKnown".equalsIgnoreCase(ip)){
        return ip;
    }
    return request.getRemoteAddr();
}


/**
 * 将xml字符串解析成map
 * @param xml
 * @return
 */
public static Map<String,String> parseXml2Map(String xml) {
    Map<String, String> map = new HashMap<String, String>();
    try {
        Document doc = DocumentHelper.parseText(xml);//将xml转为dom对象
        Element root = doc.getRootElement();//获取根节点
        Element element = root.element("xml");//获取名称为queryRequest的子节点
```

```java
            List<Element> elements = element.elements();//获取这个子节点里面的所有子元素，也可以
element.elements("userList")指定获取子元素
                for (Object obj : elements) {    //遍历子元素
                    element = (Element) obj;
                    map.put(element.getName(), element.getTextTrim());
                    System.out.println(element.getName()+"--"+element.getTextTrim());
                }
        } catch (Exception e) {
            e.printStackTrace();
        }
        return map;
    }
}
```

5. 微信小程序客户端发起最终支付，调用微信付款

```javascript
//小程序发起微信支付
wx.requestPayment({
  timeStamp: param.data.timeStamp,//timeStamp一定要是字符串类型的
  nonceStr: param.data.nonceStr,
  package: param.data.package,
  signType: 'MD5',
  paySign: param.data.paySign,
  success: function (event) {
    // success
    console.log(event);

    wx.showToast({
       title: '支付成功',
       icon: 'success',
       duration: 2000
    });
  },
  fail: function (error) {
    // fail
    console.log("支付失败")
    console.log(error)
  },
  complete: function () {
    // complete
    console.log("pay complete")
  }
});
```

6. 微信服务器进行微信支付成功后，会通知开发者后台服务器

```java
package com.xiaogang.wxpay;

import java.io.BufferedReader;
import java.io.IOException;
import java.io.InputStreamReader;
import java.util.Map;

import javax.servlet.ServletException;
import javax.servlet.ServletInputStream;
```

```java
import javax.servlet.http.HttpServlet;
import javax.servlet.http.HttpServletRequest;
import javax.servlet.http.HttpServletResponse;

import com.xiaogang.app.util.PayUtil;

/**
 * 微信服务器回调
 * @author xiaogang
 *
 */
public class WxNotifyServelt extends HttpServlet {
    private static final long serialVersionUID = 1L;


    public WxNotifyServelt() {
        super();
    }

    protected void doGet(HttpServletRequest request, HttpServletResponse response) throws ServletException, IOException {
        doPost(request, response);
    }

    protected void doPost(HttpServletRequest request, HttpServletResponse response) throws ServletException, IOException {
        BufferedReader br = new BufferedReader(new InputStreamReader((ServletInputStream)request.getInputStream()));
        String line = null;
        StringBuilder sb = new StringBuilder();
        while((line = br.readLine()) != null){
            sb.append(line);
        }
        br.close();
        //微信服务器支付成功后异步调用方式返回值
        String notityXml = sb.toString();
        String result = "fail";
        Map<String,String> map = PayUtil.parseXml2Map(notityXml);
        String returnCode = map.get("return_code");
        if("SUCCESS".equals(returnCode)){
            result = "success";
        }
        System.out.println("微信小程序支付状态："+result);
    }
}
```

4.12 分享API

微信小程序可以在Page中定义onShareAppMessage函数，设置该页面的分享信息，只有定义了此事件处理函数，右上角菜单才会显示"分享"按钮，用户单击"分享"按钮的时候会调用，此事件需要return一个Object，用于自定义分享内容，自定义分享字段如表4.59所示。

分享API

表4.59 自定义分享字段

字段	说明	默认值
title	分享标题	当前小程序名称
desc	分享描述	当前小程序名称
path	分享路径	当前页面 path，必须是以 / 开头的完整路径

示例代码如下所示。

```
Page({
  onShareAppMessage: function () {
    return {
      title: '自定义分享标题',
      desc: '自定义分享描述',
      path: '/page/user?id=123'
    }
  }
})
```

界面效果如图4.30、图4.31、图4.32所示。

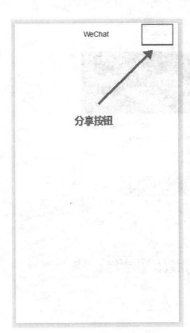

图4.30　分享按钮

图4.31　是否分享

图4.32　分享界面

4.13　沙场大练兵：仿豆瓣电影微信小程序

豆瓣电影App是一款用来购买电影票、查看影评的软件，图4.33、图4.34、图4.35、图4.36是豆瓣电影App的主要界面。

下面我们来设计一款豆瓣电影微信小程序，用它可以查看上映的电影以及电影详情等内容。

仿豆瓣电影微信小程序

图4.33 电影

图4.34 影院

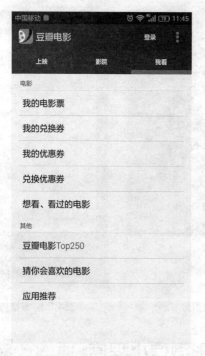

图4.35 我看

图4.36 电影详情

4.13.1 电影顶部页签切换效果

在电影界面的顶部有3个页签：上映、影院、我看。页签的切换会带动相应的内容进行切换显示，采用顶部页签切换的方式，来完成各个页面的切换显示，如图4.37所示。

（1）添加一个douban项目，填写AppID，如图4.38所示。

图4.37　顶部页签

图4.38　添加douban项目

（2）进入到app.json文件里，添加"pages/movie/movie""pages/cinema/cinema""pages/me/me""pages/movieDetail/movieDetail"4个文件路径，删除默认创建的index、logs文件路径，将窗口导航栏背景色设置为黑色（#1A1A1A），导航标题为豆瓣电影，文字颜色设置为白色（white），具体代码如下所示。

```
{
  "pages": [
    "pages/movie/movie",
    "pages/cinema/cinema",
    "pages/me/me",
    "pages/movieDetail/movieDetail"
  ],
  "window": {
    "backgroundTextStyle": "light",
    "navigationBarBackgroundColor": "#1A1A1A",
    "navigationBarTitleText": "豆瓣电影",
    "navigationBarTextStyle": "white"
  }
}
```

（3）进入到movie.wxml文件里，采用view视图容器来布局顶部的3个页签。需要设置两种样式，一种是选中样式select，另外一种是正常样式normal。定义currentTab变量与页签索引值做比较，id为索引值，添加switchNav绑定事件，具体代码如下所示。

```
<view class="nav">
  <view id="0" class="{{currentTab == 0?'select':'normal'}}" bindtap="switchNav">上映</view>
  <view class="line">|</view>
  <view id="1" class="{{currentTab == 1?'select':'normal'}}" bindtap="switchNav">影院</view>
  <view class="line">|</view>
```

```
<view id="2" class="{{currentTab == 2?'select':'normal'}}" bindtap="switchNav">我看</view>
</view>
```

（4）进入到movie.wxss文件里，给nav、select、normal、line这4个class添加样式，具体代码如下所示。

```
.nav{
    display: flex;
    flex-direction: row;
    background-color: #222222;
}
.select{
    width: 32%;
    height: 45px;
    line-height: 45px;
    text-align: center;
    color: #ffffff;
    font-size: 13px;
    border-bottom: 10px solid #777777;
}
.normal{
    width: 32%;
    height: 45px;
    line-height: 45px;
    text-align: center;
    color: #ffffff;
    font-size: 13px;
}
.line{
    height: 45px;
    line-height: 45px;
    font-size:25px;
    color: #666666;
}
```

界面效果如图4.39所示。

图4.39 顶部页签布局

（5）进入到movie.js文件里，定义currentTab变量默认值为0，添加switchNav页签单击事件，单击时获取页签的id索引值，然后赋值给currentTab变量，具体代码如下所示。

```
Page({
  data: {
    currentTab: 0
  },
  switchNav: function (e) {
    var id = e.currentTarget.id;
```

```
      this.setData({ currentTab: id });
    }
})
```
这样就可以选中顶部页签进行相应的切换显示,如图4.40、图4.41所示。

图4.40　上映页签选中　　　　　　　　　图4.41　影院页签选中

（6）进入到movie.wxml文件里,布局页签内容,单击页签切换时,页签内容滑动进行切换,需要借助swiper滑块视图组件,根据currentTab变量的索引值来判断显示swiper哪个面板的内容,具体代码如下所示。

```
<view class="nav">
  <view id="0" class="{{currentTab == 0?'select':'normal'}}" bindtap="switchNav">上映</view>
  <view class="line">|</view>
  <view id="1" class="{{currentTab == 1?'select':'normal'}}" bindtap="switchNav">影院</view>
  <view class="line">|</view>
  <view id="2" class="{{currentTab == 2?'select':'normal'}}" bindtap="switchNav">我看</view>
</view>
<swiper current="{{currentTab}}" style="height:{{winHeight}}px">
  <swiper-item>
    <view>上映内容</view>
  </swiper-item>
  <swiper-item>
    <view>影院内容</view>
  </swiper-item>
  <swiper-item>
    <view>我看内容</view>
  </swiper-item>
</swiper>
```

（7）进入到movie.js文件里,添加onLoad生命周期函数,使用wx.getSystemInfo获得窗口的高度和宽度,赋值给变量winWidth、winHeight,winWidth作为页签内容的宽度,winHeight作为页签内容的高度,具体代码如下所示。

```
Page({
  data: {
    currentTab: 0,
    winWidth: 0,
    winHeight: 0
  },
  onLoad: function (e) {
    var page = this;
    wx.getSystemInfo({
      success: function (res) {
        console.log(res);
        page.setData({ winWidth: res.windowWidth });
        page.setData({ winHeight: res.windowHeight });
```

```
      }
    })
  },
  switchNav: function (e) {
    var id = e.currentTarget.id;
    this.setData({ currentTab: id });
  }
})
```

这样就实现了顶部页签切换效果，页签切换时，页签内容也随之切换，实现了动态切换效果的显示。

4.13.2 电影海报轮播效果

海报轮播效果是很多App软件和网站都会使用到的，它可以在有限的区域内动态地展示商品图片信息或者广告信息，豆瓣电影里也有海报轮播效果，如图4.42所示。

图4.42 海报轮播区域

（1）将海报轮播的图片，复制到douban项目，进入到movie.wxml文件里，进行海报轮播效果的界面布局，具体代码如下所示。

```
<view class="nav">
  <view id="0" class="{{currentTab == 0?'select':'normal'}}" bindtap="switchNav">上映</view>
  <view class="line">|</view>
  <view id="1" class="{{currentTab == 1?'select':'normal'}}" bindtap="switchNav">影院</view>
  <view class="line">|</view>
  <view id="2" class="{{currentTab == 2?'select':'normal'}}" bindtap="switchNav">我看</view>
</view>
<swiper current="{{currentTab}}" style="height:{{winHeight}}px">
  <swiper-item>
    <view class="haibao">
      <swiper indicator-dots="{{indicatorDots}}" autoplay="{{autoplay}}" interval="{{interval}}" duration="{{duration}}" style="height:74px;">
```

```
            <block wx:for="{{imgUrls}}">
              <swiper-item>
                <image src="{{item}}" class="silde-image" style="width:100%;height:74px;"></image>
              </swiper-item>
            </block>
          </swiper>
        </view>
      </swiper-item>
      <swiper-item>
        <view>影院内容</view>
      </swiper-item>
      <swiper-item>
        <view>我看内容</view>
      </swiper-item>
</swiper>
```

（2）进入到movie.js文件里，定义海报轮播需要的变量值indicatorDots为 false、autoplay为true、interval为5000、duration为1000、imgUrls为"/images/haibao/1.jpg""/images/haibao/2.jpg""/images/haibao/3.jpg""/images/haibao/4.jpg"，具体代码如下所示。

```
Page({
  data: {
    currentTab: 0,
    winWidth: 0,
    winHeight: 0,
    indicatorDots: false,
    autoplay: true,
    interval: 5000,
    duration: 1000,
    imgUrls: [
      "/images/haibao/1.jpg",
      "/images/haibao/2.jpg",
      "/images/haibao/3.jpg",
      "/images/haibao/4.jpg"
    ]
  },
  onLoad: function (e) {
    var page = this;
    wx.getSystemInfo({
      success: function (res) {
        console.log(res);
        page.setData({ winWidth: res.windowWidth });
        page.setData({ winHeight: res.windowHeight });
      }
    })
  },
  switchNav: function (e) {
    var id = e.currentTarget.id;
    this.setData({ currentTab: id });
  }
})
```

这样就实现了海报轮播效果，界面效果如图4.43所示。

图4.43 海报轮播效果

4.13.3 电影列表方式布局

豆瓣电影的电影列表布局采用每行3列的方式来进行,来显示电影海报和电影名称,如图4.44所示。

图4.44 电影列表

(1)进入到movie.js文件里,定义loadMovies函数使用wx.request来获取豆瓣电影的电影列表信息,在onLoad函数里调用loadMovies函数。同时定义movies变量,并将电影列表赋值给movies,具体代码如下所示。

```
var util = require('../../utils/util.js')
Page({
  data: {
    currentTab: 0,
    winWidth: 0,
    winHeight: 0,
    indicatorDots: false,
    autoplay: true,
    interval: 5000,
    duration: 1000,
    imgUrls: [
      "/images/haibao/1.jpg",
```

```
      "/images/haibao/2.jpg",
      "/images/haibao/3.jpg",
      "/images/haibao/4.jpg"
    ]
  },
  onLoad: function (e) {
    var page = this;
    wx.getSystemInfo({
      success: function (res) {
        console.log(res);
        page.setData({ winWidth: res.windowWidth });
        page.setData({ winHeight: res.windowHeight });
      }
    });
    this.loadMovies();
  },
  switchNav: function (e) {
    var id = e.currentTarget.id;
    this.setData({ currentTab: id });
  },
  loadMovies: function () {
    var page = this;
    var key = util.getDataKey();
    wx.request({
      url: 'https://api.douban.com/v2/movie/in_theaters?apikey=' + key,
      method: 'GET',
      header: {
        "Content-Type": "json"
      },
      success: function (res) {
        console.log(res);
        var subjects = res.data.subjects;
        var size = subjects.length;//电影总数量
        var len = parseInt(size / 3);//每行放置3个电影, 计算出需要多少行

        console.log(len);
        console.log(subjects);
        page.setData({ movies: subjects });
        page.setData({ winHeight: (len + 1) * 230 });//动态的设置电影内容的高度

      }
    })
  }
})
```

（2）获取到电影列表信息后，进入到movie.wxml文件里，电影海报采用中等大小的图片，宽度设置为100，高度设置为150，绑定loadDetail查看电影详情事件，具体代码如下所示。

```
<view class="nav">
  <view id="0" class="{{currentTab == 0?'select':'normal'}}" bindtap="switchNav">上映</view>
  <view class="line">|</view>
  <view id="1" class="{{currentTab == 1?'select':'normal'}}" bindtap="switchNav">影院</view>
  <view class="line">|</view>
  <view id="2" class="{{currentTab == 2?'select':'normal'}}" bindtap="switchNav">我看</view>
```

```
    </view>
    <swiper current="{{currentTab}}" style="height:{{winHeight}}px">
        <swiper-item>
            <view class="haibao">
                <swiper indicator-dots="{{indicatorDots}}" autoplay="{{autoplay}}" interval="{{interval}}" duration="{{duration}}" style="height:74px;">
                    <block wx:for="{{imgUrls}}">
                        <swiper-item>
                            <image src="{{item}}" class="silde-image" style="width:100%;height:74px;"></image>
                        </swiper-item>
                    </block>
                </swiper>
            </view>
            <view class="items">
                <block wx:for="{{movies}}">
                    <view class="item" bindtap="loadDetail" id="{{item.id}}">
                        <view>
                            <image src="{{item.images.medium}}" style="width:100px;height:150px;"></image>
                        </view>
                        <view class="name">
                            <text>{{item.title}}</text>
                        </view>
                    </view>
                </block>
            </view>
        </swiper-item>
        <swiper-item>
            <view>影院内容</view>
        </swiper-item>
        <swiper-item>
            <view>我看内容</view>
        </swiper-item>
    </swiper>
```

（3）进入到movie.wxss文件里，给items、item、name这3个class添加样式，具体代码如下所示。

```
.nav{
    display: flex;
    flex-direction: row;
    background-color: #222222;
}
.select{
    width: 32%;
    height: 45px;
    line-height: 45px;
    text-align: center;
    color: #ffffff;
    font-size: 13px;
    border-bottom: 10px solid #777777;
}
.normal{
    width: 32%;
    height: 45px;
    line-height: 45px;
```

```
        text-align: center;
        color: #ffffff;
        font-size: 13px;
}
.line{
        height: 45px;
        line-height: 45px;
        font-size:25px;
        color: #666666;
}
.items{
        background-color: #f2f2f2;
        height:1000px;
}
.item{
        width:32%;
        height: 200px;
        margin-top:10px;
        text-align: center;
        float: left;
}
.name{
        font-size: 14px;
        font-weight: bold;
        margin: 5px;
}
```

这样就完成了电影列表界面布局展示,界面效果如图4.45所示。

图4.45 电影列表布局

4.13.4 电影详情页布局

在电影列表界面里，单击电影海报图片，可以查看具体的电影详情。电影详情页在顶部也是采用页签切换的方式进行布局，布局方式和电影页面一样，页签的下面是介绍电影相关信息的区域，接着是"我想看""看过了"两个按钮，再往下就是电影介绍、导演演员列表的展现，如图4.46所示。

图4.46 电影详情页

（1）进入到movie.js文件里，添加loadDetail函数，跳转到详情页里需要把电影的id带过去，详情信息需要电影id才能查得到，具体代码如下所示。

```
var util = require('../../utils/util.js')
Page({
  data: {
    currentTab: 0,
    winWidth: 0,
    winHeight: 0,
    indicatorDots: false,
    autoplay: true,
    interval: 5000,
    duration: 1000,
    imgUrls: [
      "/images/haibao/1.jpg",
      "/images/haibao/2.jpg",
      "/images/haibao/3.jpg",
      "/images/haibao/4.jpg"
    ]
  },
  onLoad: function (e) {
    var page = this;
    wx.getSystemInfo({
      success: function (res) {
        console.log(res);
        page.setData({ winWidth: res.windowWidth });
        page.setData({ winHeight: res.windowHeight });
```

```
      }
    });
    this.loadMovies();
  },
  switchNav: function (e) {
    var id = e.currentTarget.id;
    this.setData({ currentTab: id });
  },
  loadMovies: function () {
    var page = this;
    var key = util.getDataKey();
    wx.request({
      url: 'https://api.douban.com/v2/movie/in_theaters?apikey=' + key,
      method: 'GET',
      header: {
        "Content-Type": "json"
      },
      success: function (res) {
        console.log(res);
        var subjects = res.data.subjects;
        var size = subjects.length;//电影总数量
        var len = parseInt(size / 3);//每行放置3个电影，计算出需要多少行

        console.log(len);
        console.log(subjects);
        page.setData({ movies: subjects });
        page.setData({ winHeight: (len + 1) * 230 });//动态地设置电影内容的高度

      }
    })
  },
  loadDetail: function (e) {
    var id = e.currentTarget.id;
    wx.navigateTo({
      url: '../movieDetail/movieDetail?id=' + id
    })
  }
})
```

（2）进入到movieDetail.js文件里，定义3个变量：电影movie、导演directors、演员casts。在onLoad函数中，根据传递过来的id查找电影详情，然后赋值给movie、directors、casts，具体代码如下所示。

```
var util = require('../../utils/util.js')
Page({
  data: {
    currentTab: 0,
    winWidth: 0,
    winHeight: 0,
    movie: {},
    directors: [],
    casts: []
  },
  onLoad: function (e) {
    var page = this;
    var key = util.getDataKey();
    wx.request({
      url: 'https://api.douban.com/v2/movie/subject/' + e.id + '?apikey=' + key,
```

```
            header: {
                "Content-Type": "json"
            },
            success: function (res) {
                console.log(res);
                var movie = res.data;
                page.setData({ movie: movie });
                page.setData({ directors: movie.directors });
                page.setData({ casts: movie.casts });
                wx.setNavigationBarTitle({
                    title: movie.title
                })
            }
        });
    },
})
```

（3）进入到movieDetail.wxml文件里，进行顶部页签界面布局以及页签内容布局，具体代码如下所示。

```
<view class="nav">
    <view id="0" class="{{currentTab == 0?'select':'normal'}}" bindtap="switchNav">介绍</view>
    <view class="line">|</view>
    <view id="1" class="{{currentTab == 1?'select':'normal'}}" bindtap="switchNav">图片</view>
    <view class="line">|</view>
    <view id="2" class="{{currentTab == 2?'select':'normal'}}" bindtap="switchNav">短评</view>
    <view class="line">|</view>
    <view id="3" class="{{currentTab == 3?'select':'normal'}}" bindtap="switchNav">影评</view>
</view>
<swiper current="{{currentTab}}" style="height:1200px;background-color:#F9F9F9;">
    <swiper-item>
        <view>介绍内容</view>
    </swiper-item>
    <swiper-item>
        <view>图片内容</view>
    </swiper-item>
    <swiper-item>
        <view>短评内容</view>
    </swiper-item>
    <swiper-item>
        <view>影评内容</view>
    </swiper-item>
</swiper>
```

（4）进入到movieDetail.wxss文件里，给class添加相应的样式，具体代码如下所示。

```
.nav{
    display: flex;
    flex-direction: row;
    background-color: #222222;
}
.select{
    width: 32%;
    height: 30px;
    line-height: 30px;
    text-align: center;
    color: #ffffff;
    font-size: 13px;
    border-bottom: 8px solid #777777;
}
```

```css
.normal{
    width: 32%;
    height: 30px;
    line-height: 30px;
    text-align: center;
    color: #ffffff;
    font-size: 13px;
}
.line{
    height: 30px;
    line-height: 30px;
    font-size:15px;
    color: #666666;
}
```

（5）进入到movieDetail.js文件里，定义变量currentTab、winWidth、winHeight，并进行赋值，具体代码如下所示。

```js
var util = require('../../utils/util.js')
Page({
  data: {
    currentTab: 0,
    winWidth: 0,
    winHeight: 0,
    movie: {},
    directors: [],
    casts: []
  },
  onLoad: function (e) {
    var page = this;
    var key = util.getDataKey();
    wx.request({
      url: 'https://api.douban.com/v2/movie/subject/' + e.id + '?apikey=' + key,
      header: {
        "Content-Type": "json"
      },
      success: function (res) {
        console.log(res);
        var movie = res.data;
        page.setData({ movie: movie });
        page.setData({ directors: movie.directors });
        page.setData({ casts: movie.casts });
        wx.setNavigationBarTitle({
          title: movie.title
        })
      }
    });
    wx.getSystemInfo({
      success: function (res) {
        console.log(res);
        page.setData({ winWidth: res.windowWidth });
        page.setData({ winHeight: res.windowHeight });
      }
    });
  },
  switchNav: function (e) {
    var id = e.currentTarget.id;
```

```
    this.setData({ currentTab: id });
  }
})
```
界面效果如图4.47所示。

图4.47 顶部页签切换效果

（6）进入到movieDetail.wxml文件里，进行电影详情、电影简介、导演和演员的界面布局设计，具体代码如下所示。

```
<view class="nav">
  <view id="0" class="{{currentTab == 0?'select':'normal'}}" bindtap="switchNav">介绍</view>
  <view class="line">|</view>
  <view id="1" class="{{currentTab == 1?'select':'normal'}}" bindtap="switchNav">图片</view>
  <view class="line">|</view>
  <view id="2" class="{{currentTab == 2?'select':'normal'}}" bindtap="switchNav">短评</view>
  <view class="line">|</view>
  <view id="3" class="{{currentTab == 3?'select':'normal'}}" bindtap="switchNav">影评</view>
</view>
<swiper current="{{currentTab}}" style="height:1200px;background-color:#F9F9F9;">
  <swiper-item>
    <view class="movieInfo">
      <view>
        <image src="{{movie.images.medium}}" style="width:100px;height:150px"></image>
      </view>
      <view class="detail">
        <view>
          <text class="score">评分：{{movie.rating.average}}</text>（{{movie.ratings_count}}人评分）</view>
        <view>
          <text>{{movie.year}}年上映</text>
        </view>
        <view>
          <text class="desc">{{movie.genres[0]}}</text>
        </view>
        <view>
          <text class="desc">{{movie.countries[0]}}</text>
        </view>
        <view class="buy">选座购票</view>
      </view>
    </view>
    <view class="opr">
      <view>我想看</view>
      <view>看过了</view>
    </view>
    <view class="intro">
      <text>{{movie.summary}}</text>
    </view>
    <block wx:for="{{directors}}">
      <view class="personInfo">
        <view>
          <image src="{{item.avatars.small}}" style="width:70px;height:100px"></image>
```

```
            </view>
            <view class="name">
               <view>
                  <text>{{item.name}} [导演]</text>
               </view>
            </view>
         </view>
      </block>
      <block wx:for="{{casts}}">
         <view class="personInfo">
            <view>
               <image src="{{item.avatars.small}}" style="width:70px;height:100px"></image>
            </view>
            <view class="name">
               <view>
                  <text>{{item.name}}</text>
               </view>
            </view>
         </view>
      </block>
   </swiper-item>
   <swiper-item>
      <view>图片内容</view>
   </swiper-item>
   <swiper-item>
      <view>短评内容</view>
   </swiper-item>
   <swiper-item>
      <view>影评内容</view>
   </swiper-item>
</swiper>
```

（7）进入到movieDetail.wxss文件里，给class添加相应的样式，具体代码如下所示。

```
.nav{
    display: flex;
    flex-direction: row;
    background-color: #222222;
}
.select{
    width: 32%;
    height: 30px;
    line-height: 30px;
    text-align: center;
    color: #ffffff;
    font-size: 13px;
    border-bottom: 8px solid #777777;
}
.normal{
    width: 32%;
    height: 30px;
    line-height: 30px;
    text-align: center;
    color: #ffffff;
    font-size: 13px;
}
.line{
    height: 30px;
```

```
    line-height: 30px;
    font-size:15px;
    color: #666666;
}
.movieInfo{
    display: flex;
    flex-direction:row;
    margin: 15px;
}
.detail{
    margin-left:15px;
    font-size: 13px;
}
.detail view{
    margin-bottom: 7px;
}
.score{
   color:#FC7F60;
}
.desc{
    font-weight: bold;
}
.buy{
    width: 155px;
    height:40px;
    line-height: 40px;
    background-color: #2DADDC;
    text-align: center;
    color: #ffffff;
}
.opr{
    display: flex;
    flex-direction: row;
}
.opr view{
    width: 45%;
    height: 40px;
    line-height: 40px;
    margin: 0 auto;
    background-color: #EEEEEE;
    font-size: 13px;
    text-align: center;
    border-radius: 3px;
}
.intro{
    background-color: #F1F1F1;
    margin: 10px;
    font-size: 13px;
    padding: 10px;
    line-height: 15px;
}
.personInfo{
    display: flex;
    flex-direction:row;
    margin: 15px;
    background-color: #F1F1F1;
}
```

```
.name{
    font-size: 14px;
    font-weight: bold;
    line-height: 100px;
    margin-left: 10px;
}
```

这样就实现了电影详情页界面的设计,界面效果如图4.48所示。

图4.48　电影详情页界面

豆瓣电影微信小程序主要实现了页签的切换效果、海报轮播效果、豆瓣电影列表的展现、电影详情页的设计,需要用到view视图内容组件、image图片组件、swiper滑块视图组件等组件,同时用到了wx.request、wx.getSystemInfo等API接口,并进行网络通信获取json数据,将分析到的json数据展现在页面里。

4.13.5　项目上传与预览

项目开发完成后,可以上传到微信小程序服务器上,如图4.49所示。

图4.49　项目上传

可以在手机上预览豆瓣电影微信小程序，如图4.50所示。

预览的时候，需要扫描二维码，才能在手机上进行预览，预览效果如图4.51、图4.52所示。

图4.50　项目预览（本图中二维码只是示意，请扫描自己操作生成的二维码）

图4.51　电影列表

图4.52　电影详情

4.14　小结

本章主要学习微信小程序API的使用，重点应该掌握以下内容：

（1）掌握微信小程序如何请求服务器数据；

（2）掌握微信小程序文件上传、下载和WebSocket会话API的使用；

（3）掌握微信小程序图片处理、文件操作、数据缓存API的使用；

（4）了解微信小程序位置信息、设备应用API的使用；

（5）掌握微信小程序交互反馈、登录、微信支付、分享API的使用。

第5章
微信小程序设计及问答

- 微信小程序设计
- 微信小程序问答
- 小结

设计微信小程序需要注意在一些细节上提升交互效果,营造清晰流畅的用户体验。本章介绍一些微信小程序设计时的注意事项和经常遇到的问题。

微信小程序设计及问答

5.1 微信小程序设计

5.1.1 突出重点，减少干扰项

每个页面都应有明确的重点，以便于用户每进入一个新页面的时候都能快速地理解页面内容。在确定了重点的情况下，应尽量避免页面上出现其他与用户的决策和操作无关的干扰因素。

反面示例如图5.1所示。

正确示例如图5.2所示。

图5.1 干扰项过多

图5.2 减少干扰项

5.1.2 主次动作区分明显

在一个界面上有多个按钮的时候，按钮设计要有主次之分，并且区分明显，让用户看到后就知道他能做什么、该怎么做。

反面示例如图5.3所示。

正确示例如图5.4所示。

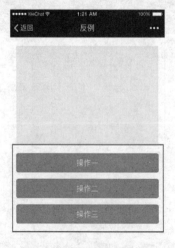

图5.3 按钮没有主次之分

图5.4 按钮有主次之分

5.1.3 流程明确，避免打断

当用户在进入某个页面进行某一个操作流程时，应避免出现用户目标流程之外的内容而打断用户。

例如：用户想进入某个页面购买商品，突然弹出抽奖的模态窗口界面，等用户抽完奖之后，可能就会忘记去买商品这件事，这对我们引导用户购买商品很不利，所有要尽量避免打断用户的主要流程操作，如图5.5所示就是用抽奖打断用户操作的界面设计。

图5.5 抽奖打断用户操作

5.1.4 局部加载反馈

局部加载反馈即只在触发加载的页面局部进行反馈，这样的反馈机制更加有针对性，页面跳动小，是微信推荐的反馈方式，如图5.6所示。

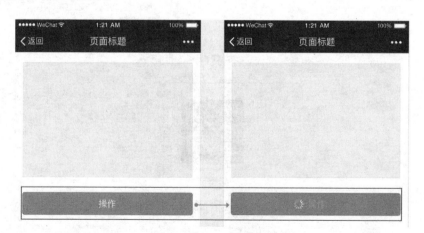

图5.6 局部加载反馈

5.1.5 模态窗口加载反馈

模态的加载样式会覆盖整个页面，由于无法明确告知具体加载的位置或内容将可能引起用户的焦虑感，因此应谨慎使用该样式，在某些全局性操作下不要使用模态的加载，如图5.7所示。

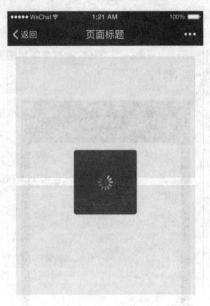

图5.7　模态窗口加载反馈

5.1.6　弹出式操作结果

弹出式提示（Toast）适用于轻量级的成功提示，它1.5秒后自动消失，并不打断流程，对用户影响较小，适用于不需要强调的操作提醒，例如成功提示。特别注意该形式不适用于错误提示，因为错误提示需明确告知用户，因而不适合使用一闪而过的弹出式提示，如图5.8所示。

图5.8　弹出式操作结果

5.1.7　模态对话框操作结果

对于需要用户明确知晓的操作结果状态可通过模态对话框来提示，并可附带下一步操作指引，如图5.9所示。

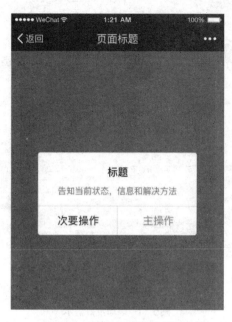

图5.9 模态对话框操作结果

5.1.8 结果页

对于操作结果已经是当前流程的终结的情况，可使用操作结果页来反馈。这种方式最为强烈和明确地告知用户操作已经完成，并可根据实际情况给出下一步操作的指引，如图5.10所示。

图5.10 结果页

5.1.9 表单填写友好提示

用户在填写表单时，输入格式或者内容不符合表单填写规则，需要给用户及时反馈表单填写问题，可以在表单顶部告知错误原因，并标识出错误字段提示用户修改，如图5.11所示。

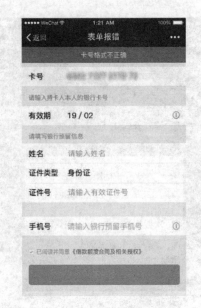

图5.11　表单友好提示

5.2　微信小程序问答

1. 如何将元素固定在界面，不随着界面滚动？

界面底部有4个导航菜单：筛选、出发时间、旅行时间、显示价格。把它们固定在界面底部，如图5.12所示。

图5.12　固定在界面底部

wxml示例代码如下所示：

```
<view class="bottomNav">
    <view id="0" class="common" bindtap="switchNav">筛选</view>
    <view style="color:#ffffff">|</view>
```

```
<view id="1" class="common" bindtap="switchNav">出发时间</view>
<view style="color:#ffffff">|</view>
<view id="2" class="common" bindtap="switchNav">旅行时间</view>
<view style="color:#ffffff">|</view>
<view id="3" class="common" bindtap="switchNav">显示价格</view>
</view>
```

wxss示例代码如下所示:

```
.bottomNav{
    background-color: #505963;
    display: flex;
    flex-direction: row;
    height: 45px;
    line-height: 45px;
    position: fixed;
    bottom:0px;
    width: 100%;
}
.bottomNav view{
    margin: 0 auto;
}
.common{
    font-size: 13px;
    color: #ffffff;
}
```

2. 怎么样获取用户在表单组件输入的内容?

能够获取用户输入的内容，需要使用组件的属性bindchange将用户的输入内容同步到 AppService。

```
<input id="myInput" bindchange="bindChange" />
<checkbox id="myCheckbox" bindchange="bindChange" />

var inputContent = {}

Page({
  data: {
    inputContent: {}
  },
  bindChange: function(e) {
    inputContent[e.currentTarget.id] = e.detail.value
  }
})
```

3. 为什么脚本内不能使用window等对象?

页面的脚本逻辑是在JsCore中运行的，JsCore是一个没有窗口对象的环境，所以不能在脚本中使用window，也无法在脚本中操作组件。

4. wx.navigateTo无法同时打开超过5个页面?

一个应用只能同时打开5个页面，当已经打开了5个页面之后，wx.navigateTo不能正常打开新页面。请避免多层级的交互方式，或者使用wx.redirectTo函数。

5. 如何修改窗口的背景色?

使用 page 标签选择器，可以修改顶层节点的样式。

```
page {
  display: block;
```

```
    min-height: 100%;
    background-color: red;
}
```

6. 如何跳转的时候带参数和跳转到的界面接收参数？

跳转带参数，示例代码如下所示：

```
Page({
  btn: function () {
    wx.navigateTo({
      url: '../../index/index?id=' + 1000 + "&name=" + kevin
    })
  }
})
```

接收参数，示例代码如下所示：

```
Page({
  onLoad:function(e){
    var id = e.id;
    var name = e.name;
  }
})
```

5.3 小结

本章主要介绍微信小程序在设计过程中遇到的问题以及如何提高用户的体验度，重点掌握以下内容：

（1）在设计过程中要突出重点，减少干扰项，给用户明确的主次操作，让用户操作流程更顺畅；

（2）用户在操作过程中要及时反馈，局部加载以及模态窗口加载都是对用户的操作反馈；

（3）在用户操作后要给出其明确的操作结果，可以通过弹出式操作结果、模态对话框操作结果等告知用户；

（4）用户在填写表单时，要进行友好的提示和正确的引导，减少用户填写表单的时间以及出错的概率；

（5）理解微信小程序设计过程中遇到的一些问题以及解决方案。

第二篇 综合案例应用

第6章

综合案例：仿智行火车票12306微信小程序

智行火车票是一款收取费用的自动查询预订火车票的软件，可以实时监控余票的多少，与铁路部门数据实时同步。该软件可以直接在12306上订票，可以查询、预定和购买车票。本章案例模仿制作智行火车票小程序。

- 需求描述
- 设计思路及相关知识点
- 准备工作
- 设计流程
- 小结

智行火车票对12306购票流程进行了大量优化,使用户购票更加快捷,还额外提供了智能查询和火车票监控功能。智行火车票App的主要界面如图6.1、图6.2、图6.3、图6.4所示。

图6.1　火车票

图6.2　飞机票

图6.3　汽车票

图6.4　个人中心

6.1 需求描述

仿智行火车票12306微信小程序要完成以下功能。

（1）底部标签导航，放置4个标签：火车票、飞机票、汽车票、个人中心。选中效果图片呈现为蓝色，字体颜色设置蓝色，如图6.5所示。

图6.5 标签选中效果

（2）在火车票界面，顶部区域放置海报轮播区域，中间放置火车票、飞机票查询内容，下面放置快捷导航菜单：极速抢票、在线选座、抢手好货、超值酒店。如图6.6所示。

（3）在火车票界面，输入起始站和终点站，单击查询按钮，可以查到火车票信息，包括车站名称、车次、历时时长、发车时间、到站时间、票价，如图6.7所示。

图6.6 火车票界面

图6.7 火车票列表

（4）在个人中心界面，设计账号信息内容、订单菜单导航以及二级页面的列表式导航，如图6.8所示。

（5）在个人中心界面，单击"邀请好友一起来抢票"导航，可以查看抢票情况，如图6.9所示。

图6.8 个人中心界面　　　　　图6.9 抢票界面

6.2 设计思路及相关知识点

6.2.1 设计思路

（1）设计底部标签导航，准备好底部标签导航的图标并建立相应的4个页面；设置默认时图片和选中时图片，标签名称采用两种颜色，蓝色为选中颜色，灰色为默认颜色。

（2）设计幻灯片轮播效果，准备好幻灯片需要轮播的图片。

（3）设计火车票查询区域，火车票查询区域有火车票和飞机票两个页签的切换效果，切换选中时，背景色为白色，文字颜色为黑色。

（4）设计火车票列表界面时，可以先设计出一条火车票信息，然后发起网络请求获得所有的火车票列表，采用列表渲染的方式展现出所有火车票信息。

（5）个人中心采用列表式导航的方式来进行二级界面导航，这种导航设计也是先设计出一个菜单，其他的导航菜单可以直接复用这个菜单的界面效果。

（6）抢票界面也是非常有规律性的界面，先设计出第一个区域内容，其他区域内容可以直接进行复用。

6.2.2 相关知识点

（1）在界面布局的时候，会用到微信小程序的组件，包括view视图容器组件、image图片组件、swiper滑块视图容器组件、表单相关组件等组件的使用。

（2）界面样式设计，需要写一些wxss样式进行界面的美化和渲染。

（3）获取火车票列表信息，需要使用wx.request发起网络请求获取到火车票相关信息，返回json数据，在界面中进行渲染。

（4）界面跳转需要使用wx.navigateTo这个API接口来实现。

6.3 准备工作

（1）需要准备一个AppID，如果没有AppID，就不能在手机上进行项目的预览，但是在开发工具上开发是没有任何问题的。

（2）底部标签导航，需要有选中图片和默认图片，放置在images/bar文件夹下，如图6.10所示。

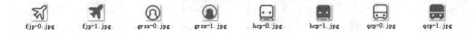

图6.10　底部标签导航图片

（3）需要准备海报轮播的图片，放置在images/haibao文件夹下，如图6.11所示。

图6.11　海报轮播图片

（4）准备火车票界面用到的一些图标，放置在images/icon/hcp文件夹下，如图6.12所示。

图6.12　火车票界面图标

（5）准备个人中心界面用到的一些图标，放置在images/icon/grzx文件夹下，如图6.13所示。

图6.13　个人中心界面图标

6.4 设计流程

仿智行火车票12306微信小程序的设计流程为：先设计底部标签导航，添加导航对应的4个界面，即火车票、飞机票、汽车票、个人中心；在火车票界面里设计海报轮播效果、火车票查询界面；再添加一个新的火车票列表界面，在这个界面里完成火车票列表设计；在个人中心界面里完成个人中心界面设计；最后添加一个抢票界面，完成抢票界面的设计。

设计流程

6.4.1　底部标签导航设计

仿智行火车票12306微信小程序的底部标签导航分为4个标签导航：火车票、飞机票、汽车票、个人中心。标签导航选中时导航图标会变为蓝色图标，导航文字会变为蓝色文字，如图6.14所示。

图6.14　底部标签导航选中效果

（1）新建一个zxtrain项目的微信小程序，将准备好的底部标签导航图标、海报轮播图片、火车票界面图标、个人中心界面图标放置在zxtrain项目下。

（2）打开app.json配置文件，在pages数组里添加4个页面路径"pages/train/train""pages/airplane/airplane""pages/bus/bus""pages/mycenter/mycenter"，保存后会自动生成相应的页面文件夹；删除"pages/index/index""pages/logs/logs"页面路径以及对应的文件夹，具体代码如下所示。

```
{
  "pages":[
    "pages/train/train",
    "pages/airplane/airplane",
    "pages/bus/bus",
    "pages/mycenter/mycenter"
  ],
  "window":{
    "backgroundTextStyle":"light",
    "navigationBarBackgroundColor": "#fff",
    "navigationBarTitleText": "WeChat",
    "navigationBarTextStyle":"black"
  }
}
```

（3）在window数组里配置窗口导航背景颜色为蓝色（#5495E6），导航栏文字为智行12306，字体颜色设置为白色（#ffffff），具体代码如下所示。

```
{
  "pages":[
    "pages/train/train",
    "pages/airplane/airplane",
    "pages/bus/bus",
    "pages/mycenter/mycenter"
  ],
  "window":{
    "backgroundTextStyle":"light",
    "navigationBarBackgroundColor": "#5495E6",
    "navigationBarTitleText": "智行12306",
    "navigationBarTextStyle":"white"
  }
}
```

（4）在tabBar对象里配置底部标签导航背景色为白色（#ffffff），文字默认颜色为灰色，选中时为蓝色（#5495E6），在list数组里配置底部标签导航对应的页面、导航名称、默认时图标、选中时图标，具体代码如下所示。

```
{
  "pages":[

    "pages/train/train",
    "pages/airplane/airplane",
    "pages/bus/bus",
    "pages/mycenter/mycenter",
    "pages/trainList/trainList",
    "pages/grabticket/grabticket"

  ],
  "window":{
    "backgroundTextStyle":"light",
    "navigationBarBackgroundColor": "#5495E6",
    "navigationBarTitleText": "智行12306",
    "navigationBarTextStyle":"white"
  },
  "tabBar": {
    "selectedColor": "#5495E6",
    "backgroundColor": "#ffffff",
    "borderStyle": "white",
    "list": [{
      "pagePath": "pages/train/train",
      "text": "火车票",
      "iconPath": "images/bar/hcp-0.jpg",
      "selectedIconPath": "images/bar/hcp-1.jpg"
    },{
      "pagePath": "pages/airplane/airplane",
      "text": "飞机票",
      "iconPath": "images/bar/fjp-0.jpg",
      "selectedIconPath": "images/bar/fjp-1.jpg"
    },{
      "pagePath": "pages/bus/bus",
      "text": "汽车票",
      "iconPath": "images/bar/qcp-0.jpg",
      "selectedIconPath": "images/bar/qcp-1.jpg"
    },{
      "pagePath": "pages/mycenter/mycenter",
      "text": "个人中心",
      "iconPath": "images/bar/grzx-0.jpg",
      "selectedIconPath": "images/bar/grzx-1.jpg"
    }]
  }
}
```

这样就完成了仿智行火车票12306微信小程序的底部标签导航配置，单击不同的导航，可以切换显示不同的页面，同时导航图标和导航文字会呈现为选中状态，如图6.15所示。

图6.15　火车票界面

6.4.2　海报轮播效果设计

海报轮播效果可以在有限的区域内动态地显示不同的幻灯片图片,是很多网站或者App软件都会采用的一种展现方式,在仿智行火车票12306微信小程序的火车票界面里,采用海报轮播效果展示广告图片,如图6.16所示。

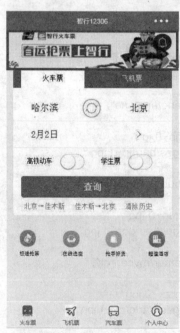

图6.16　海报轮播显示

(1)进入到pages/train/train.wxml文件里,采用view、swiper、image进行布局,图片宽度设置为100%,高度设置为80px,具体代码如下所示:

```
<view class="haibao">
    <swiper indicator-dots="{{indicatorDots}}" autoplay="{{autoplay}}" interval="{{interval}}"
```

```
duration="{{duration}}" style="height:80px;">
    <block wx:for="{{imgUrls}}">
      <swiper-item>
        <image src="{{item}}" style="width:100%;height:80px;"></image>
      </swiper-item>
    </block>
  </swiper>
</view>
```

swiper滑块视图容器设置为自动播放(autoplay="true")，自动切换时间间隔为3s(interval="3000")，滑动动画时长为1s(duration="1000");

采用wx:for循环来显示要展示的图片，从train.js文件里获取imgUrls图片路径。

（2）进入到pages/ train/train.js文件中，在data对象里定义imgUrls数组，存放海报轮播的图片路径，代码如下：

```
Page({
  data:{
    indicatorDots:false,
    autoplay:true,
    interval:5000,
    duration:1000,
    imgUrls:[
      '/images/haibao/1.jpg',
      '/images/haibao/2.jpg',
      '/images/haibao/3.jpg'
    ]
  },
  onLoad:function(options){
    // 页面初始化 options为页面跳转所带来的参数
  }
})
```

（3）这样就可以实现幻灯片轮播效果，如图6.17、图6.18所示。

图6.17 海报轮播一

图6.18 海报轮播二

6.4.3 火车票查询界面设计

火车票查询界面可以输入始发站、终点站、出行日期、火车类型等内容来进行火车票查询；提供飞机票页签，其与火车票页签可以进行相互切换显示；在火车票查询界面下面是4个快捷导航菜单：极速抢票、在线选座、抢手好货、超值酒店。如图6.19所示。

图6.19 火车票界面

1. 页签切换效果设计

（1）进入到pages/train/train.wxml文件里，设计火车票与飞机票页签切换效果以及选中状态效果，设计两种样式：一种是选中样式select，另一种是正常样式normal。根据变量currentTab值来决定使用哪个样式，同时提供switchNav切换导航的事件，具体代码如下所示。

```
<view class="haibao">
    <swiper indicator-dots="{{indicatorDots}}" autoplay="{{autoplay}}" interval="{{interval}}" duration="{{duration}}" style="height:80px;">
        <block wx:for="{{imgUrls}}">
            <swiper-item>
                <image src="{{item}}" style="width:100%;height:80px;"></image>
            </swiper-item>
        </block>
    </swiper>
</view>
<view class="content">
    <view class="navbg">
        <view id="0" class="{{currentTab == 0?'select':'normal'}}" bindtap="switchNav">火车票</view>
        <view id="1" class="{{currentTab == 1?'select':'normal'}}" bindtap="switchNav">飞机票</view>
    </view>
</view>
```

（2）进入到pages/train/train.wxss文件里，添加导航背景灰色（#898989），灰色背景上面是圆角矩形;添加页签选中时和默认时的样式,页签选中时背景色是白色(#ffffff),文字是黑色(#000000),页签默认时文字颜色是白色(#ffffff),具体代码如下所示。

```css
.content{
    height:500px;
    background-color: #F4F4F4;
}
.navbg{
    width: 92%;
    background-color: #898989;
    height: 40px;
    margin: 0 auto;
    border-top-left-radius:5px;
    border-top-right-radius:5px;
    display: flex;
    flex-direction: row;
}
.select{
    width: 40%;
    height: 40px;
    line-height: 40px;
    text-align: center;
    color: #000000;
    font-size: 15px;
    margin: 0 auto;
    background-color: #ffffff;
}
.normal{
    width: 40%;
    height: 40px;
    line-height: 40px;
    text-align: center;
    color: #ffffff;
    font-size: 15px;
    margin: 0 auto;
}
```

(3)进入到pages/train/train.js文件里,定义变量currentTab默认值为0,添加switchNav事件,用来进行页签相互切换,动态改变变量currentTab的值,具体代码如下所示。

```js
Page({
  data:{
        indicatorDots:false,
        autoplay:true,
        interval:5000,
        duration:1000,
        imgUrls:[
            '/images/haibao/1.jpg',
            '/images/haibao/2.jpg',
            '/images/haibao/3.jpg'
        ],
        currentTab:0
  },
  onLoad:function(options){
        // 页面初始化 options为页面跳转所带来的参数
  },
```

```
    switchNav:function (e) {
      var id = e.currentTarget.id;
      this.setData({ currentTab: id });
    }
})
```
界面效果如图6.20所示。

图6.20　页签切换效果

2．火车票查询区域设计

（1）进入到pages/train/train.wxml文件里，设计火车票查询区域，采用表单组件input文本框来输入始发站、终点站、日期，采用switch组件来选择高铁动车和学生票，采用button组件提交按钮，采用form组件来提交表单，具体代码如下所示。

```
<view class="haibao">
  <swiper indicator-dots="{{indicatorDots}}" autoplay="{{autoplay}}" interval="{{interval}}" duration="{{duration}}" style="height:80px;">
    <block wx:for="{{imgUrls}}">
      <swiper-item>
        <image src="{{item}}" style="width:100%;height:80px;"></image>
      </swiper-item>
    </block>
  </swiper>
</view>
<view class="content">
  <view class="navbg">
    <view id="0" class="{{currentTab == 0?'select':'normal'}}" bindtap="switchNav">火车票</view>
    <view id="1" class="{{currentTab == 1?'select':'normal'}}" bindtap="switchNav">飞机票</view>
  </view>
  <view class="formbg">
    <form bindsubmit="formSubmit">
      <view class="station">
        <view>
          <input name="startStation" value="哈尔滨" />
```

```
          </view>
          <view>
            <image src="../../images/icon/hcp/xz.jpg" style="width:44px;height:45px;"></image>
          </view>
          <view>
            <input name="endStation" value="北京" />
          </view>
        </view>
        <view class="hr"></view>
        <view class="station">
          <view>
            <input name="date" value="2月2日" />
          </view>
          <view></view>
          <view>
            <text style="color:#5495E6;">
              <input name="week" value="2月2日" />
            </text>></view>
        </view>
        <view class="hr"></view>
        <view class="type">
          <view>高铁动车
            <switch name="gt" type="switch" />
          </view>
          <view>学生票
            <switch name="xs" type="switch" />
          </view>
        </view>
        <button class="btn" formType="submit">查询</button>
        <view class="record">
          <text>北京→佳木斯</text>
          <text>佳木斯→北京</text>
          <text>清除历史</text>
        </view>
      </form>
  </view>
</view>
```

（2）进入到pages/train/train.wxss文件里，添加相应的样式，具体代码如下所示。

```
.content{
    height:500px;
    background-color: #F4F4F4;
}
.navbg{
    width: 92%;
    background-color: #898989;
    height: 40px;
    margin: 0 auto;
    border-top-left-radius:5px;
    border-top-right-radius:5px;
    display: flex;
    flex-direction: row;
}
.select{
    width: 40%;
```

```css
        height: 40px;
        line-height: 40px;
        text-align: center;
        color: #000000;
        font-size: 15px;
        margin: 0 auto;
        background-color: #ffffff;
    }
    .normal{
        width: 40%;
        height: 40px;
        line-height: 40px;
        text-align: center;
        color: #ffffff;
        font-size: 15px;
        margin: 0 auto;
    }
    .formbg{
        width: 92%;
        background-color: #ffffff;
        margin: 0 auto;
        padding-top:20px;
        padding-bottom: 10px;
        border-bottom-left-radius: 5px;
        border-bottom-right-radius: 5px;
    }
    .station{
        display: flex;
        flex-direction: row;
        width: 90%;
        margin: 0 auto;
        text-align: center;
    }
    .station view{
        height: 45px;
        line-height: 45px;
        font-size: 20px;
        width: 33%;
    }
    .station input{
       height: 45px;
       line-height: 45px;
    }
    .hr{
        height: 1px;
        background-color: #cccccc;
        opacity: 0.2;
        margin-top: 5px;
        margin-bottom: 5px;
    }
    .type{
        display: flex;
        flex-direction: row;
        width: 90%;
        margin: 0 auto;
```

```
        text-align: center;
}
.type view{
        height: 45px;
        line-height: 45px;
        font-size: 14px;
        width: 50%;
}
.type switch{
        margin-left:10px;
}
.btn{
        width: 90%;
        height: 45px;
        line-height: 45px;
        color: #ffffff;
        text-align: center;
        font-size: 20px;
        background-color: #5495E6;
        margin: 0 auto;
        margin-top:10px;
        border-radius: 5px;
}
.record{
        text-align: center;
        margin-top:10px;
        font-size: 15px;
        color: #999999;
}
.record text{
        margin-right: 20px;
}
```

界面效果如图6.21所示。

图6.21　火车票查询表单

（3）在app.json文件中配置一个新的页面路径"pages/trainList/trainList"，用来设计火车票列表界面，微信小程序框架会自动建立相应的trainList文件夹。

（4）在pages/train/train.js文件里，添加表单提交formSubmit事件，获得始发站、终点站、日期、星期数据值，把这些数据值带到trainList火车票列表界面，具体代码如下所示。

```js
Page({
  data:{
    indicatorDots:false,
    autoplay:true,
    interval:5000,
    duration:1000,
    imgUrls:[
      '/images/haibao/1.jpg',
      '/images/haibao/2.jpg',
      '/images/haibao/3.jpg'
    ],
    currentTab:0
  },
  onLoad:function(options){
    // 页面初始化 options为页面跳转所带来的参数
  },
  switchNav:function (e) {
    var id = e.currentTarget.id;
    this.setData({ currentTab: id });
  },
  formSubmit:function(e){
    console.log(e);
    var startStation = e.detail.value.startStation;//始发站
    var endStation = e.detail.value.endStation;//终点站
    var date = e.detail.value.date;//日期：2月2日
    var week = e.detail.value.week;//星期：周四
    wx.navigateTo({
      url: '../trainList/trainList?startStation=' + startStation+"&endStation="+endStation+"&date="+date+"&week="+week
    })
  }
})
```

在这个火车票查询界面里，输入始发站和终点站，单击查询按钮，就会根据输入的内容进行相应的火车票列表的查询。

3. 快捷导航设计

（1）进入到pages/train/train.wxml文件里，在火车票查询下面有4个查询按钮，它是由图标和导航名称组成的，具体代码如下所示。

```html
<view class="haibao">
  <swiper indicator-dots="{{indicatorDots}}" autoplay="{{autoplay}}" interval="{{interval}}" duration="{{duration}}" style="height:80px;">
    <block wx:for="{{imgUrls}}">
      <swiper-item>
        <image src="{{item}}" style="width:100%;height:80px;"></image>
      </swiper-item>
    </block>
  </swiper>
</view>
<view class="content"
```

```
<view class="navbg">
  <view id="0" class="{{currentTab == 0?'select':'normal'}}" bindtap="switchNav">火车票</view>
  <view id="1" class="{{currentTab == 1?'select':'normal'}}" bindtap="switchNav">飞机票</view>
</view>
<view class="formbg">
  <form bindsubmit="formSubmit">
    <view class="station">
      <view>
        <input name="startStation" value="哈尔滨" />
      </view>
      <view>
        <image src="../../images/icon/hcp/xz.jpg" style="width:44px;height:45px;"></image>
      </view>
      <view>
        <input name="endStation" value="北京" />
      </view>
    </view>
    <view class="hr"></view>
    <view class="station">
      <view>
        <input name="date" value="2月2日" />
      </view>
      <view></view>
      <view>
        <text style="color:#5495E6;">
          <input name="week" value="2月2日" />
        </text></view>
    </view>
    <view class="hr"></view>
    <view class="type">
      <view>高铁动车
        <switch name="gt" type="switch" />
      </view>
      <view>学生票
        <switch name="xs" type="switch" />
      </view>
    </view>
    <button class="btn" formType="submit">查询</button>
    <view class="record">
      <text>北京→佳木斯</text>
      <text>佳木斯→北京</text>
      <text>清除历史</text>
    </view>
  </form>
</view>
  <view class="icon">
  <view class="item">
    <view>
      <image src="../../images/icon/hcp/jsqp.jpg" style="width:40px;height:40px;"></image>
    </view>
    <view class="menu">极速抢票</view>
  </view>
  <view class="item">
    <view>
      <image src="../../images/icon/hcp/zxxz.jpg" style="width:40px;height:40px;"></image>
```

```
        </view>
        <view class="menu">在线选座</view>
      </view>
      <view class="item">
        <view>
          <image src="../../images/icon/hcp/qshh.jpg" style="width:40px;height:40px;"></image>
        </view>
        <view class="menu">抢手好货</view>
      </view>
      <view class="item">
        <view>
          <image src="../../images/icon/hcp/czjd.jpg" style="width:40px;height:40px;"></image>
        </view>
        <view class="menu">超值酒店</view>
      </view>
    </view>
</view>
```

（2）进入到pages/train/train.wxss文件里，给icon、item、menu这3个class添加样式，具体代码如下所示。

```
.content{
    height:500px;
    background-color: #F4F4F4;
}
.navbg{
    width: 92%;
    background-color: #898989;
    height: 40px;
    margin: 0 auto;
    border-top-left-radius:5px;
    border-top-right-radius:5px;
    display: flex;
    flex-direction: row;
}
.select{
    width: 40%;
    height: 40px;
    line-height: 40px;
    text-align: center;
    color: #000000;
    font-size: 15px;
    margin: 0 auto;
    background-color: #ffffff;
}
.normal{
    width: 40%;
    height: 40px;
    line-height: 40px;
    text-align: center;
    color: #ffffff;
    font-size: 15px;
    margin: 0 auto;
}
.formbg{
    width: 92%;
```

```css
    background-color: #ffffff;
    margin: 0 auto;
    padding-top:20px;
    padding-bottom: 10px;
    border-bottom-left-radius: 5px;
    border-bottom-right-radius: 5px;
}
.station{
    display: flex;
    flex-direction: row;
    width: 90%;
    margin: 0 auto;
    text-align: center;
}
.station view{
    height: 45px;
    line-height: 45px;
    font-size: 20px;
    width: 33%;
}
.station input{
    height: 45px;
    line-height: 45px;
}
.hr{
    height: 1px;
    background-color: #cccccc;
    opacity: 0.2;
    margin-top: 5px;
    margin-bottom: 5px;
}
.type{
    display: flex;
    flex-direction: row;
    width: 90%;
    margin: 0 auto;
    text-align: center;
}
.type view{
    height: 45px;
    line-height: 45px;
    font-size: 14px;
    width: 50%;
}
.type switch{
    margin-left:10px;
}
.btn{
    width: 90%;
    height: 45px;
    line-height: 45px;
    color: #ffffff;
    text-align: center;
    font-size: 20px;
    background-color: #5495E6;
```

```
    margin: 0 auto;
    margin-top:10px;
    border-radius: 5px;
}
.record{
    text-align: center;
    margin-top:10px;
    font-size: 15px;
    color: #999999;
}
.record text{
    margin-right: 20px;
}
.icon{
    display: flex;
    flex-direction: row;
    margin-top: 30px;
}
.item{
    width: 25%;
    text-align: center;
    margin: 0 auto;
}
.menu{
    font-size: 11px;
    width:100px;
}
```

界面效果如图6.22所示。

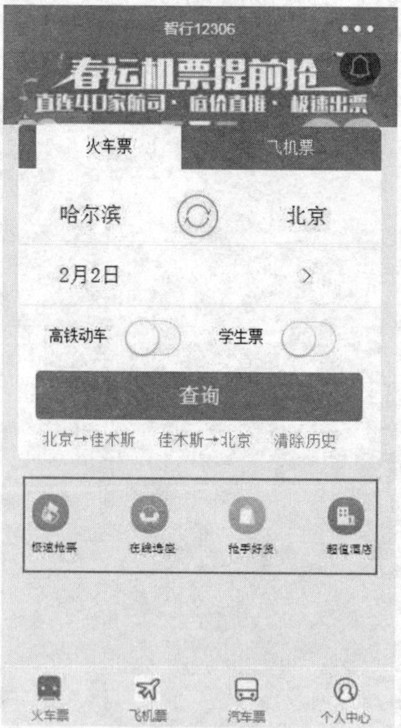

图6.22 快捷导航设计

这样就完成了火车票和飞机票页签的切换效果、火车票查询表单设计以及快捷导航设计，输入表单内容，就可以根据这些表单内容进行火车票列表查询。

6.4.4 火车票列表设计

从火车票查询界面跳转到火车票列表界面，会把始发站、终点站、日期这些查询条件带到火车票列表界面中，根据这些查询条件，会加载出相应的火车票信息，如图6.23所示。

1. 导航标题和日期展示

在火车票列表的最上面是导航标题和日期的展示，导航标题是从上一个界面传递过来的始发站和终点站的参数，日期也是从上一个界面传递过来的参数，如图6.24所示。

图6.23　火车票列表界面

图6.24　导航标题和日期

（1）进入到pages/trainList/trainList.js文件里，在onLoad生命周期函数里接收上一个界面传递过来的参数，并通过wx.setNavigationBarTitle来设置导航标题，具体代码如下所示。

```
Page({
  data:{
  },
  onLoad:function(e){
    var startStation = e.startStation;//始发站
    var endStation = e.endStation;//终点站
    var date = e.date;//日期
    console.log("startStation="+startStation+"---endStation="+endStation+"---date="+date);
    wx.setNavigationBarTitle({
      title: startStation+'→'+endStation
    });
  }
})
```

（2）进入到pages/trainList/trainList.wxml文件里，进行日期设计，包括查询的日期、前一天和后一天，具体代码如下所示。

```
<view class="date">
    <view>前一天</view>
    <view>02月02日周四</view>
    <view>后一天</view>
</view>
```

（3）进入到pages/trainList/trainList.wxss文件里，给date这个class添加样式，设计成蓝色背景，文字设置成白色（#ffffff），具体代码如下所示。

```
.date{
    height:40px;
    background-color: #5495E6;
    display: flex;
    flex-direction: row;
}
.date view{
    margin: 0 auto;
    color: #ffffff;
    padding-top: 10px;
}
```

（4）进入到pages/trainList/trainList.js文件里，定义一个日期date变量，把传递过来的日期赋值给date变量，具体代码如下所示。

```
Page({
  data:{
    date:"
  },
  onLoad:function(e){
    var startStation = e.startStation;//始发站
    var endStation = e.endStation;//终点站
    var date = e.date;//日期
    console.log("startStation="+startStation+"---endStation="+endStation+"---date="+date);
    wx.setNavigationBarTitle({
      title: startStation+'→'+endStation
    });
    this.setData({data:date});
  }
})
```

（5）进入到pages/trainList/trainList.wxml文件里，将date变量绑定到wxml文件里，动态显示日期，具体代码如下所示。

```
<view class="date">
    <view>前一天</view>
    <view>{{date}}</view>
    <view>后一天</view>
</view>
```

界面效果如图6.25所示。

图6.25 动态显示日期

2. 火车票列表设计

火车票列表是很有规律地进行展示的列表，每一条火车票信息都包括始发站、终点站、车次、日期、票价等各种信息。先来设计一条火车票信息，然后去请求火车票列表接口，来获取火车票列表信息，最后用列表渲染的方式展现出来。

（1）进入到pages/trainList/trainList.wxml文件里，设计一条火车票信息的布局，具体代码如下所示。

```
<view class="date">
  <view>前一天</view>
  <view>{{date}}</view>
  <view>后一天</view>
</view>
<view class="content" style="height:{{winHeight}}px">
  <view class="bg">
    <view class="item">
      <view class="wrApper left">
        <view class="normal">哈尔滨</view>
        <view class="blue">北京</view>
      </view>
      <view class="wrApper center">
        <view class="normal">D28</view>
        <view class="line"></view>
        <view class="small">7小时54分</view>
      </view>
      <view class="wrApper right">
        <view class="normal">06:57</view>
        <view class="normal">14:51</view>
      </view>
      <view class="wrApper right">
        <view class="blue">￥306.5起</view>
        <view class="buy">可抢票</view>
      </view>
    </view>
    <view class="hr"></view>
    <view class="seat">
      <view class="yes">一等座:10张<text>(抢)</text></view>
      <view class="no">二等座:0张<text>(抢)</text></view>
    </view>
  </view>
</view>
```

（2）进入到pages/trainList/trainList.wxss文件里，给火车票信息添加样式，具体代码如下所示。

```
.date{
    height:40px;
    background-color: #5495E6;
    display: flex;
    flex-direction: row;
}
.date view{
    margin: 0 auto;
    color: #ffffff;
    padding-top: 10px;
}
.content{
    height:600px;
```

```css
    background-color: #F4F4F4;
    padding-top:10px;
}
.bg{
    width: 95%;
    height: 90px;
    background-color: #ffffff;
    margin: 0 auto;
    border-radius: 5px;
    margin-bottom: 10px;
}
.item{
    display: flex;
    flex-direction: row;
    padding: 10px;

}
.wrApper{
    width: 25%;
}
.left{
    text-align: left;
}
.center{
    text-align: center;
}
.right{
    text-align: right;
}
.blue{
    color: #5495E6;
    font-weight: bold;
    font-size: 16px;
}
.normal{
    color: #000000;
    font-weight: bold;
    font-size: 16px;
}
.small{
    font-size: 13px;
    color: #666666;
}
.line{
    height: 1px;
    background-color: #cccccc;
    opacity: 0.2;
    width: 80%;
    margin: 0 auto;
    margin-top:3px;
    margin-bottom:3px;
}
.buy{
    background-color: red;
```

```
        width: 42px;
        height: 20px;
        line-height: 20px;
        font-size: 12px;
        color: #ffffff;
        text-align: center;
        float: right;
        border-radius: 20px;
        margin-right: 5px;
    }
    .hr{
        height: 1px;
        background-color: #cccccc;
        opacity: 0.2;
    }
    .seat{
        font-size: 13px;
        display: flex;
        flex-direction: row;
        margin-top: 5px;
        margin-left: 10px;
    }
    .seat text{
        color: red;
    }
    .no{
        color: #999999;
        margin-right:10px;
    }
    .yes{
        margin-right:10px;
    }
```

界面效果如图6.26所示。

图6.26 火车票信息

（3）进入到pages/trainList/trainList.js文件里，去动态获取火车票列表信息，定义trainList变量来绑定火车票列表信息，定义winHeight变量来动态计算列表信息，具体代码如下所示。

```
var util = require('../../utils/util.js')
Page({
  data:{
    date:'',
    trainList:[],
    winHeight:600,
    currentTab:'1'
```

```javascript
},
onLoad:function(e){
    var startStation = e.startStation;//始发站
    var endStation = e.endStation;//终点站
    var date = e.date;//日期
    console.log("startStation="+startStation+"---endStation="+endStation+"---date="+date);
    wx.setNavigationBarTitle({
        title: startStation+'→'+endStation
    });
    this.setData({date:date});
    this.loadTrainsList(startStation,endStation);
},
loadTrainsList:function(startStation,endStation){
    var page = this;
    var key = util.getDataKey();
    console.log(key)
    wx.request({
        url: 'https://api.apishop.net/common/train/getLeftTicket?apiKey=' + key + '&date=2018-09-05' + '&startStation=' + startStation + '&endStation=' + endStation,
        method: 'GET',
        success: function(res){
            console.log(res);
            var trainList = res.data.result;
            console.log(trainList);
            var size = trainList.length;
            var winHeight = size * 100 + 30;
            page.setData({trainList:trainList});
            page.setData({winHeight:winHeight});
        }
    });
}
})
```

trainList数据结构如下所示。

```
▼ {data: {…}, header: {…}, statusCode: 200, errMsg: "request:ok"}
  ▼ data:
       desc: "请求成功"
     ▼ result: Array(19)
       ▼ 0:
            costtime: "19:58"
            day: "1"
            departstation: "鸡西"
            ed: "--"
            endstation: "北京"
            endtime: "21:26"
            gr: "--"
            qt: "--"
            rw: "无"
            rz: "--"
            starttime: "01:28"
            station: "哈尔滨东"
            sw: "--"
            td: "--"
            terminalstation: "北京"
            trainno: "K40"
            type: "快速"
            wz: "有"
            yd: "--"
            yw: "无"
            yz: "有"
          ▶ __proto__: Object
```

火车票数据接口有请求次数限制的,一旦请求次数到达限制,就不能返回数据,可以去apishop网站自己注册账号,免费获取数据接口。

(4)进入到pages/trainList/trainList.wxml文件里,将trainList变量、winHeight变量动态地绑定到界面里,具体代码如下所示。

```
<view class="date">
  <view>前一天</view>
  <view>{{date}}</view>
  <view>后一天</view>
</view>
<view class="content" style="height:{{winHeight}}px">
  <block wx:for="{{trainList}}">
    <view class="bg">
      <view class="item">
        <view class="wrapper left">
          <view class="normal">{{item.station}}</view>
          <view class="blue">{{item.starttime}}</view>
        </view>
        <view class="wrapper center">
          <view class="normal">{{item.trainno}}</view>
          <view class="line"></view>
          <view class="small">{{item.costtime}}</view>
        </view>
        <view class="wrapper right">
          <view class="normal">{{item.terminalstation}}</view>
          <view class="normal">{{item.endtime}}</view>
        </view>
        <view class="wrapper right">
          <view class="blue">￥300起</view>
          <view class="buy">可抢票</view>
        </view>
      </view>
      <view class="hr"></view>
      <view class="seat">
        <view class="yes">二等座:100张
          <text>(抢)</text>
        </view>
        <view class="yes">一等座:20张
          <text>(抢)</text>
        </view>
        <view>
          <view class="no">商务座:0张
            <text>(抢)</text>
          </view>
        </view>
      </view>
    </view>
  </block>
</view>
```

界面效果如图6.27所示。

图6.27　火车票列表

设计列表或者有规律布局的界面时，我们可以先设计共有的区域内容作为基础单元，然后复制或者列表循环展示这块共有的内容，其方法就像火车票列表信息一样。

3．火车票底部固定页签导航

火车票列表界面最底部区域，是一块固定区域的页签导航，它不随着界面的滚动而滚动，而是固定在底部，它有4个页签导航：筛选、出发时间、旅行时间、显示价格。根据这些页签导航可以进行火车票列表的展示。

（1）进入到pages/trainList/trainList.wxml文件里，设计底部固定导航，需要设计两种样式：一种是导航标题选中效果select，文字呈现为蓝色；另一种是默认样式效果common，定义变量currentTab来动态渲染样式，定义switchNav来切换事件，具体代码如下所示。

```
<view class="date">
  <view>前一天</view>
  <view>{{date}}</view>
  <view>后一天</view>
</view>
<view class="content" style="height:{{winHeight}}px">
  <block wx:for="{{trainList}}">
    <view class="bg">
      <view class="item">
        <view class="wrapper left">
          <view class="normal">{{item.station}}</view>
          <view class="blue">{{item.starttime}}</view>
        </view>
        <view class="wrapper center">
          <view class="normal">{{item.trainno}}</view>
          <view class="line"></view>
```

```
            <view class="small">{{item.costtime}}</view>
          </view>
          <view class="wrapper right">
            <view class="normal">{{item.terminalstation}}</view>
            <view class="normal">{{item.endtime}}</view>
          </view>
          <view class="wrapper right">
            <view class="blue">￥300起</view>
            <view class="buy">可抢票</view>
          </view>
        </view>
        <view class="hr"></view>
        <view class="seat">
          <view class="yes">二等座:100张
            <text>(抢)</text>
          </view>
          <view class="yes">一等座:20张
            <text>(抢)</text>
          </view>
          <view>
            <view class="no">商务座:0张
              <text>(抢)</text>
            </view>
          </view>
        </view>
      </view>
    </block>
    <view class="bottomNav">
      <view id="0" class="{{currentTab==0?'selected':'common'}}" bindtap="switchNav">筛选</view>
      <view style="color:#ffffff">|</view>
      <view id="1" class="{{currentTab==1?'selected':'common'}}" bindtap="switchNav">出发时间</view>
      <view style="color:#ffffff">|</view>
      <view id="2" class="{{currentTab==2?'selected':'common'}}" bindtap="switchNav">旅行时间</view>
      <view style="color:#ffffff">|</view>
      <view id="3" class="{{currentTab==3?'selected':'common'}}" bindtap="switchNav">显示价格</view>
    </view>
</view>
```

（2）进入到pages/trainList/trainList.wxss文件里，添加页签导航选中效果样式和默认效果样式，具体代码如下所示。

```
.date{
    height:40px;
    background-color: #5495E6;
    display: flex;
    flex-direction: row;
}
.date view{
    margin: 0 auto;
    color: #ffffff;
    padding-top: 10px;
}
.content{
```

```
        height:600px;
        background-color: #F4F4F4;
        padding-top:10px;
}
.bg{
        width: 95%;
        height: 90px;
        background-color: #ffffff;
        margin: 0 auto;
        border-radius: 5px;
        margin-bottom: 10px;
}
.item{
        display: flex;
        flex-direction: row;
        padding: 10px;

}
.wrApper{
        width: 25%;
}
.left{
        text-align: left;
}
.center{
        text-align: center;
}
.right{
        text-align: right;
}
.blue{
        color: #5495E6;
        font-weight: bold;
        font-size: 16px;
}
.normal{
        color: #000000;
        font-weight: bold;
        font-size: 16px;
}
.small{
        font-size: 13px;
        color: #666666;
}
.line{
        height: 1px;
        background-color: #cccccc;
        opacity: 0.2;
        width: 80%;
        margin: 0 auto;
        margin-top:3px;
        margin-bottom:3px;
```

```css
}
.buy{
    background-color: red;
    width: 42px;
    height: 20px;
    line-height: 20px;
    font-size: 12px;
    color: #ffffff;
    text-align: center;
    float: right;
    border-radius: 20px;
    margin-right: 5px;
}
.hr{
    height: 1px;
    background-color: #cccccc;
    opacity: 0.2;
}
.seat{
    font-size: 13px;
    display: flex;
    flex-direction: row;
    margin-top: 5px;
    margin-left: 10px;
}
.seat text{
    color: red;
}
.no{
    color: #999999;
    margin-right:10px;
}
.yes{
    margin-right:10px;
}
.bottomNav{
    background-color: #505963;
    display: flex;
    flex-direction: row;
    height: 45px;
    line-height: 45px;
    position: fixed;
    bottom:0px;
    width: 100%;
}
.bottomNav view{
    margin: 0 auto;
}
.selected{
    font-size: 13px;
    color: #5495E6;
}
```

```
.common{
    font-size: 13px;
    color: #ffffff;
}
```

(3)进入到pages/trainList/trainList.js文件里,定义变量currentTab值为1,添加菜单切换事件switchNav,用来动态设置页签导航选中效果,具体代码如下所示。

```
var util = require('../../utils/util.js')
Page({
  data:{
    date:'',
    trainList:[],
    winHeight:600,
    currentTab:'1'
  },
  onLoad:function(e){
    var startStation = e.startStation;//始发站
    var endStation = e.endStation;//终点站
    var date = e.date;//日期
    console.log("startStation="+startStation+"---endStation="+endStation+"---date="+date);
    wx.setNavigationBarTitle({
      title: startStation+'→'+endStation
    });
    this.setData({date:date});
    this.loadTrainsList(startStation,endStation);
  },
  loadTrainsList:function(startStation,endStation){
    var page = this;
    var key = util.getDataKey();
    console.log(key)
    wx.request({
      url: 'https://api.apishop.net/common/train/getLeftTicket?apiKey=' + key + '&date=2018-09-05' + '&startStation=' + startStation + '&endStation=' + endStation,
      method: 'GET',
      success: function(res){
        console.log(res);
        var trainList = res.data.result;
        console.log(trainList);
        var size = trainList.length;
        var winHeight = size * 100 + 30;
        page.setData({trainList:trainList});
        page.setData({winHeight:winHeight});
      }
    });
  },
  switchNav:function (e) {
    var id = e.currentTarget.id;
    console.log(id);
    this.setData({ currentTab: id });
  }
})
```

界面效果如图6.28所示。

图6.28 固定页签导航

6.4.5 个人中心界面设计

个人中心界面用来显示账号相关信息、订单情况、我的财富、出行服务、邀请好友、消息中心、产品意见等内容,它是通过列表式导航的方式来设计界面的,如图6.29所示。

图6.29 个人中心界面

1. 账号信息设计

(1)进入到pages/mycenter/mycenter.wxml文件里,设计账号的头像、昵称等信息,使用变量userInfo作为个人信息来展示,具体代码如下所示。

```
<view class="amountBg">
    <view class="img"><image src="../../images/icon/grzx/tx.jpg" style="width:49px;height:47px;"></image></view>
    <view class="account">
        <view>{{userInfo.nickName}}</view>
        <view>账号管理</view>
    </view>
    <view class="nav">></view>
</view>
```

（2）进入到pages/mycenter/mycenter.wxss文件里，给头像、账号、二级界面导航入口添加样式，具体代码如下所示。

```
.amountBg{
    height:100px;
    background-color: #5495E6;
    display: flex;
    flex-direction: row;
    align-items: center;
}
.img{
    margin-left: 20px;
}
.account{
    width: 70%;
    color: #ffffff;
    margin-left: 10px;
}
.nav{
    width:15px;
    color: #ffffff;
}
```

（3）进入到pages/mycenter/mycenter.js文件里，动态获取昵称，使用App.getUserInfo接口来获取个人信息，具体代码如下所示。

```
var App = getApp()
Page({
    data:{
        userInfo: {}
    },
    onLoad:function(options){
        var that = this
        //调用应用实例的方法获取全局数据
        App.getUserInfo( function(userInfo) {
            console.log(userInfo);
            //更新数据
            that.setData( {
                userInfo: userInfo
            })
        })
    }
})
```

界面效果如图6.30所示。

2. 订单导航设计

（1）进入到pages/mycenter/mycenter.wxml文件里，设计火车票订单、抢票订单、机票订单、全部订单导航，它们是由图标和文字组成的，具体代码如下所示。

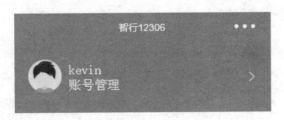

图6.30 账号信息

```
<view class="amountBg">
  <view class="img">
    <image src="../../images/icon/grzx/tx.jpg" style="width:49px;height:47px;"></image>
  </view>
  <view class="account">
    <view>{{userInfo.nickName}}</view>
    <view>账号管理</view>
  </view>
  <view class="nav">></view>
</view>
<view class="content">
  <view class="order">
    <view class="desc">
      <view><image src="../../images/icon/grzx/hcpdd.jpg" style="width:22px;height:25px;"></image></view>
      <view>火车票订单</view>
    </view>
    <view class="desc">
      <view><image src="../../images/icon/grzx/qpdd.jpg" style="width:22px;height:25px;"></image></view>
      <view>抢票订单</view>
    </view>
    <view class="desc">
      <view><image src="../../images/icon/grzx/jpdd.jpg" style="width:22px;height:25px;"></image></view>
      <view>机票订单</view>
    </view>
    <view class="desc">
      <view><image src="../../images/icon/grzx/qbdd.jpg" style="width:22px;height:25px;"></image></view>
      <view>全部订单</view>
    </view>
  </view>
</view>
```

（2）进入到pages/mycenter/mycenter.wxss文件里，给订单区域添加样式，具体代码如下所示。

```
.amountBg{
    height:100px;
    background-color: #5495E6;
    display: flex;
    flex-direction: row;
    align-items: center;
}
.img{
    margin-left: 20px;
```

```
}
.account{
    width: 70%;
    color: #ffffff;
    margin-left: 10px;
}
.nav{
    width:15px;
    color: #ffffff;
}
.content{
    background-color: #F4F4F4;
    height: 500px;
}
.order{
    width: 94%;
    height: 70px;
    display: flex;
    flex-direction: row;
    background-color: #ffffff;
    border-radius: 5px;
    text-align: center;
    align-items: center;
    position: absolute;
    top:90px;
    margin-left: 3%;
}
.desc{
    width: 25%;
    font-size: 13px;
}
```

界面效果如图6.31所示。

图6.31　订单信息

3. 列表导航设计

（1）进入到pages/mycenter/mycenter.wxml文件里，设计我的财富、出行服务、邀请好友、消息中心、产品意见、更多这6个列表导航，具体代码如下所示。

```
<view class="amountBg">
  <view class="img">
    <image src="../../images/icon/grzx/tx.jpg" style="width:49px;height:47px;"></image>
  </view>
  <view class="account">
    <view>{{userInfo.nickName}}</view>
    <view>账号管理</view>
  </view>
  <view class="nav">></view>
</view>
<view class="content">
  <view class="order">
    <view class="desc">
      <view><image src="../../images/icon/grzx/hcpdd.jpg" style="width:22px;height:25px;"></image></view>
      <view>火车票订单</view>
    </view>
    <view class="desc">
      <view><image src="../../images/icon/grzx/qpdd.jpg" style="width:22px;height:25px;"></image></view>
      <view>抢票订单</view>
    </view>
    <view class="desc">
      <view><image src="../../images/icon/grzx/jpdd.jpg" style="width:22px;height:25px;"></image></view>
      <view>机票订单</view>
    </view>
    <view class="desc">
      <view><image src="../../images/icon/grzx/qbdd.jpg" style="width:22px;height:25px;"></image></view>
      <view>全部订单</view>
    </view>
  </view>
  <view class="clear"></view>
  <view class="item">
    <view class="icon"><image src="../../images/icon/grzx/wdcf.jpg" style="width:22px;height:21px;"></image></view>
    <view class="itemName">
      <view>我的财富</view>
      <view class="remark">0个加速币</view>
    </view>
    <view class="right"><text class="opr">领取加速包</text></view>
  </view>
  <view class="line"></view>
  <view class="item">
    <view class="icon"><image src="../../images/icon/grzx/cxfw.jpg" style="width:22px;height:21px;"></image></view>
    <view class="itemName">
      <view>出行服务</view>
    </view>
    <view class="right"><text class="opr">正晚点/时刻表</text></view>
  </view>
```

```
        <view class="hr"></view>
          <view class="item">
    <view class="icon"><image src="../../images/icon/grzx/yqhy.jpg" style="width:22px;height:21px;">
    </image></view>
        <view class="itemName">
            <view>邀请好友</view>
        </view>
        <view class="right" bindtap="grabTicket"><text class="opr">一起来抢票</text></view>
     </view>
       <view class="hr"></view>
          <view class="item">
    <view class="icon"><image src="../../images/icon/grzx/xxzx.jpg" style="width:22px;height:21px;">
    </image></view>
        <view class="itemName">
            <view>消息中心</view>
        </view>
        <view class="right"><text class="opr">在线服务</text></view>
     </view>
       <view class="hr"></view>
          <view class="item">
    <view class="icon"><image src="../../images/icon/grzx/cpyj.jpg" style="width:22px;height:21px;">
    </image></view>
        <view class="itemName">
            <view>产品意见</view>
        </view>
        <view class="right"></view>
     </view>
       <view class="hr"></view>
          <view class="item">
    <view class="icon"><image src="../../images/icon/grzx/gd.jpg" style="width:22px;height:21px;">
    </image></view>
        <view class="itemName">
            <view>更多</view>
        </view>
        <view class="right"></view>
     </view>
   </view>
```

（2）进入到pages/mycenter/mycenter.wxss文件里，给我的财富、出行服务、邀请好友、消息中心、产品意见、更多这6个列表导航添加样式，具体代码如下所示。

```
.amountBg{
    height:100px;
    background-color: #5495E6;
    display: flex;
    flex-direction: row;
    align-items: center;
}
.img{
    margin-left: 20px;
}
.account{
    width: 70%;
```

```css
        color: #ffffff;
        margin-left: 10px;
}
.nav{
    width:15px;
    color: #ffffff;
}
.content{
    background-color: #F4F4F4;
    height: 500px;
}
.order{
    width: 94%;
    height: 70px;
    display: flex;
    flex-direction: row;
    background-color: #ffffff;
    border-radius: 5px;
    text-align: center;
    align-items: center;
    position: absolute;
    top:90px;
    margin-left: 3%;
}
.desc{
    width: 25%;
    font-size: 13px;
}
.clear{
    padding-top: 70px;
}
.item{
    background-color: #ffffff;
    display: flex;
    flex-direction: row;
    height: 50px;
    align-items: center;
}
.icon{
    width: 50px;
    text-align: center;
}
.itemName{
    width: 40%;
    font-size: 14px;
    font-weight: bold;
}
.line{
    height: 10px;
}
.hr{
    height: 1px;
```

```
        background-color: #cccccc;
        opacity: 0.2;
}
.remark{
    font-weight: normal;
    margin-top:5px;
}
.right{
    width: 40%;
    text-align: right;
}
.opr{
    color: #5495E6;
    font-size: 13px;
    font-weight: bold;
    margin-right: 10px;
}
```

界面效果如图6.32所示。

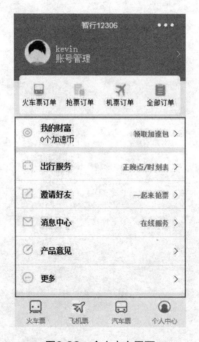

图6.32 个人中心界面

（3）给邀请好友添加grabTicket抢票事件，单击这个导航会进入到抢票界面查看自己的抢票情况，进入到pages/mycenter/mycenter.js文件里，绑定grabTicket事件，让它跳转到grabticket界面，在app.json文件里配置grabticket路径，具体代码如下所示。

```
var App = getApp()
Page({
  data:{
    userInfo: {}
  },
  onLoad:function(options){
    var that = this
```

```
        //调用应用实例的方法获取全局数据
        App.getUserInfo( function(userInfo) {
          console.log(userInfo);
          //更新数据
          that.setData( {
            userInfo: userInfo
          })
        })
      },
      grabTicket:function(){
        wx.navigateTo({
          url: '../grabticket/grabticket'
        })
      }
    })
```

通过列表导航进入到二级界面，其实现方式和进入到抢票界面一样，绑定跳转事件，配置相应的跳转路径，就可以进入到二级界面。

6.4.6 抢票界面设计

抢票界面分为3部分内容：第1部分是告诉用户抢到票后会以短信或者电话的方式进行通知；第2部分是抢票情况；第3部分是分享内容，如图6.33所示。

图6.33 抢票界面

1. 通知区域设计

（1）进入到pages/grabticket/grabticket.wxml文件里，设计抢票通知区域内容，包括攻略、抢票通知信息、二级界面入口，具体代码如下所示。

```
<view class="amountBg">
    <view class="bg">
        <view class="icon">攻略</view>
        <view class="tip">抢票成功后以电话或短信通知，收到后请及时支付！</view>
        <view class="right">></view>
    </view>
</view>
```

（2）进入到pages/ grabticket /grabticket.wxss文件里，给通知区域添加样式，设置为蓝色背景，在蓝色背景上面添加一块矩形区域用来放置通知区域内容，具体代码如下所示。

```
.amountBg{
    height:70px;
    background-color: #5495E6;
}
.bg{
    width: 80%;
    height: 53px;
    background-color: #ffffff;
    border-radius: 5px;
    margin: 0 auto;
    display: flex;
    flex-direction: row;
    align-items: center;
}
.icon{
    border: 1px solid #6EBAFE;
    font-size: 13px;
    margin-left: 10px;
    color:#6EBAFE;
    text-align: center;
}
.tip{
    font-size: 15px;
    margin: 5px;
}
.right{
    margin-right: 5px;
}
```

界面效果如图6.34所示。

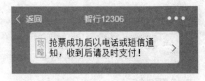

图6.34 通知区域

2. 抢票信息设计

（1）进入到pages/ grabticket /grabticket.wxml文件里，设计已经取消或者重新开始的抢票情况，具体代码如下所示。

```
<view class="amountBg">
    <view class="bg">
```

```
        <view class="icon">攻略</view>
        <view class="tip">抢票成功后以电话或短信通知，收到后请及时支付！</view>
        <view class="right">>></view>
      </view>
   </view>
<view class="content">
<view class="hr"></view>
   <view class="item">
      <view class="ticket">
         <view class="station">哈尔滨→北京</view>
         <view class="desc1">02月03日</view>
         <view class="desc2">已取消</view>
      </view>
      <view class="opr">
         查看
      </view>
   </view>
<view class="hr"></view>
<view class="item">
   <view class="ticket">
      <view class="station">北京→佳木斯</view>
      <view class="desc1">02月07日</view>
      <view class="desc2">已取消</view>
   </view>
   <view class="opr">
      查看
   </view>
</view>
<view class="hr"></view>
<view class="item">
   <view class="ticket">
      <view class="station">北京→南京</view>
      <view class="desc1">02月06日</view>
      <view class="desc2">已取消</view>
   </view>
   <view class="start">
      开始
   </view>
</view>
<view class="hr"></view>
<view class="item">
   <view class="ticket">
      <view class="station">合肥→北京</view>
      <view class="desc1">02月07日</view>
      <view class="desc2">抢票终止</view>
   </view>
   <view class="start">
      开始
   </view>
</view>
</view>
```

（2）进入到pages/ grabticket /grabticket.wxss文件里，添加相应的样式，具体代码如下所示。

```css
.amountBg{
    height:70px;
    background-color: #5495E6;
}
.bg{
    width: 80%;
    height: 53px;
    background-color: #ffffff;
    border-radius: 5px;
    margin: 0 auto;
    display: flex;
    flex-direction: row;
    align-items: center;
}
.icon{
    border: 1px solid #6EBAFE;
    font-size: 13px;
    margin-left: 10px;
    color:#6EBAFE;
    text-align: center;
}
.tip{
    font-size: 15px;
    margin: 5px;
}
.right{
    margin-right: 5px;
}
.content{
    background-color: #F4F4F4;
    height: 600px;
}
.hr{
    height: 10px;
}
.item{
    background-color: #ffffff;
    width: 90%;
    margin: 0 auto;
    padding:10px;
    display: flex;
    flex-direction: row;
    align-items: center;
}
.station{
    font-size: 15px;
    font-weight: bold;
    margin-bottom:5px;
}
.desc1{
    font-size:15px;
    color: #999999;
```

```css
        margin-bottom:15px;
}
.desc2{
    font-size:13px;
    color: #999999;
}
.ticket{
    width: 85%;
}
.opr{
    width: 45px;
    height: 45px;
    border-radius: 50px;
    background-color: #29CE73;
    color: #ffffff;
    line-height: 45px;
    font-size: 13px;
    text-align: center;
}

.start{
    width: 45px;
    height: 45px;
    border-radius: 50px;
    border: 1px solid red;
    color: red;
    line-height: 45px;
    font-size: 13px;
    text-align: center;
}
```
界面效果如图6.35所示。

图6.35　抢票信息

3. 分享信息设计

（1）进入到pages/ grabticket /grabticket.wxml文件里，设计分享区域内容，包括加速按钮、分享赢加速包以及进入到二级界面的入口，具体代码如下所示。

```
<view class="amountBg">
    <view class="bg">
        <view class="icon">攻略</view>
        <view class="tip">抢票成功后以电话或短信通知，收到后请及时支付！</view>
        <view class="right">></view>
    </view>
</view>
<view class="content">
<view class="hr"></view>
    <view class="item">
        <view class="ticket">
            <view class="station">哈尔滨→北京</view>
            <view class="desc1">02月03日</view>
            <view class="desc2">已取消</view>
        </view>
        <view class="opr">
            查看
        </view>
    </view>
<view class="hr"></view>
    <view class="item">
        <view class="ticket">
            <view class="station">北京→佳木斯</view>
            <view class="desc1">02月07日</view>
            <view class="desc2">已取消</view>
        </view>
        <view class="opr">
            查看
        </view>
    </view>
<view class="hr"></view>
    <view class="item">
        <view class="ticket">
            <view class="station">北京→南京</view>
            <view class="desc1">02月06日</view>
            <view class="desc2">已取消</view>
        </view>
        <view class="start">
            开始
        </view>
    </view>
<view class="hr"></view>
    <view class="item">
        <view class="ticket">
            <view class="station">合肥→北京</view>
            <view class="desc1">02月07日</view>
            <view class="desc2">抢票终止</view>
        </view>
```

```
            <view class="start">
                开始
            </view>
        </view>
    </view>
</view>
<view class="share">
        <view class="speed">加速</view>
        <view class="tipShare">
            <view>分享赢加速包</view>
            <view class="jsb">分享给好友，可随机赢取最多10个加速包</view>
        </view>
        <view class="detail">></view>
    </view>
```

（2）进入到pages/ grabticket /grabticket.wxss文件里，给分享区域内容添加样式，设计加速按钮，具体代码如下所示。

```
.amountBg{
    height:70px;
    background-color: #5495E6;
}
.bg{
    width: 80%;
    height: 53px;
    background-color: #ffffff;
    border-radius: 5px;
    margin: 0 auto;
    display: flex;
    flex-direction: row;
    align-items: center;
}
.icon{
    border: 1px solid #6EBAFE;
    font-size: 13px;
    margin-left: 10px;
    color:#6EBAFE;
    text-align: center;
}
.tip{
    font-size: 15px;
    margin: 5px;
}
.right{
    margin-right: 5px;
}
.content{
    background-color: #F4F4F4;
    height: 600px;
}
.hr{
    height: 10px;
}
.item{
```

```
        background-color: #ffffff;
        width: 90%;
        margin: 0 auto;
        padding:10px;
        display: flex;
        flex-direction: row;
        align-items: center;
    }
    .station{
        font-size: 15px;
        font-weight: bold;
        margin-bottom:5px;
    }
    .desc1{
        font-size:15px;
        color: #999999;
        margin-bottom:15px;
    }
    .desc2{
        font-size: 13px;
        color: #999999;
    }
    .ticket{
        width: 85%;
    }
    .opr{
        width: 45px;
        height: 45px;
        border-radius: 50px;
        background-color: #29CE73;
        color: #ffffff;
        line-height: 45px;
        font-size: 13px;
        text-align: center;
    }

    .start{
        width: 45px;
        height: 45px;
        border-radius: 50px;
        border: 1px solid red;
        color: red;
        line-height: 45px;
        font-size: 13px;
        text-align: center;
    }
    .share{
        background-color: #ffffff;
        display: flex;
        flex-direction: row;
        position: fixed;
        bottom:0px;
```

```css
    width: 100%;
    align-items: center;
    height: 55px;
}
.speed{
     width: 32px;
    height: 32px;
    border-radius: 50px;
    background-color: #29CE73;
    font-size: 11px;
    color: #ffffff;
    text-align: center;
    line-height: 32px;
    margin-left: 10px;
}
.tipShare{
    width: 80%;
}
.jsb{
    color: #999999;
    font-size: 13px;
    margin-top:5px;
}
.detail{
    width: 15px;
    text-align: right;
}
```

界面效果如图6.36所示。

图6.36 分享区域

这样就完成了抢票界面的设计，通过界面的布局设计和样式设计，就可以实现抢票界面的显示。如果想修改窗口标题，只需要在pages/ grabticket /grabticket.json文件里配置"navigationBarTitleText"的"抢票"

属性，就可以覆盖app.json文件里这个属性。

6.4.7 项目上传和预览

仿智行火车票12306微信小程序设计完成后，如果该项目有AppID，可以将微信小程序上传到微信服务器上，扫描二维码获得管理员同意后就可以上传项目，如图6.37所示。

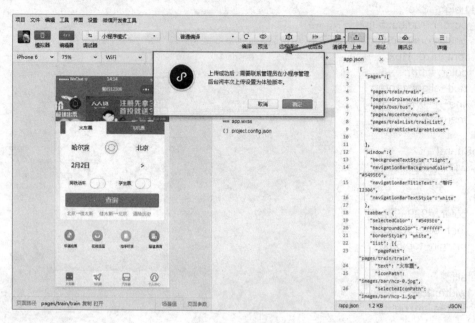

图6.37 项目上传

除了项目上传，也可以进行项目预览，在手机上运行微信小程序，同样需要扫描二维码进行项目预览，如图6.38所示。

图6.38 项目预览（图中二维码只是示意，请扫描自己操作生成的二维码）

6.5 小结

本章主要设计了仿智行火车票12306微信小程序，重点掌握以下内容：

（1）掌握底部标签导航配置、顶部页签切换效果设计，通过不同页签之间的切换，向用户展示动态的内容，实现不同内容的展示；

（2）掌握表单组件如何提交表单内容，这些表单内容如何传递到其他界面，以及其他界面如何获取传递的内容；

（3）掌握界面的布局以及给界面布局添加相应的样式和绑定事件；

（4）学会将js文件里的数据动态地绑定到WXML界面，实现数据的动态绑定；

（5）学会如何设计列表内容或者列表导航界面，通过先设计一个基本的内容区域，然后复用这个区域内容的方式，来提高开发效率；

（6）学会wx.request、wx.navigateTo、wx.setNavigationBarTitle等开发接口的使用，了解每个API不同的用处和可以实现的不同效果。

第7章
综合案例：仿糗事百科微信小程序

- 需求描述
- 设计思路及相关知识点
- 准备工作
- 设计流程
- 小结

糗事百科App，是以糗友的真实糗事为主题的笑话App，它的话题轻松休闲，在年轻人中十分流行。本章案例模仿制作糗事百科小程序。

第7章 综合案例：仿糗事百科微信小程序

在糗事百科中可以查看他人发布的糗事并与网友分享自己亲身经历或听说到的各类生活糗事，如图7.1、图7.2所示。

图7.1 专享　　　　　　　　　图7.2 视频

7.1 需求描述

仿糗事百科微信小程序主要完成以下功能。

（1）实现顶部页签菜单左右滑动效果，如图7.3所示。

图7.3 顶部页签

需求描述、设计思路及相关知识点

（2）实现顶部页签菜单切换效果，页签菜单选中时字体加粗，同时对应的内容也相应变化，如图7.4、图7.5所示。

（3）实现专享界面糗事列表设计，包括发布人头像、发布人昵称、发布的段子等信息，以列表的形式展现出来。

（4）实现视频列表页设计，使视频可以进行播放与暂停。

（5）实现分享功能，可以将当前界面分享给好友，如图7.6所示。

图7.4　专享界面　　　　图7.5　视频界面　　　　图7.6　分享页面

7.2　设计思路及相关知识点

7.2.1　设计思路

（1）实现顶部页签滑动效果，需要借助于scroll-view可滚动视图区域组件，设置scroll-x="true"属性，允许顶部页签在水平方向上左右滑动；

（2）页签菜单切换，内容也会随着进行切换，需要使用swiper滑块视图容器组件，根据current当前页面索引值来决定显示哪个面板；

（3）设计糗事列表，首先设计一条内容，然后复制这条内容的布局，在这个基础上进行修改；

（4）设计视频列表，需要使用video视频组件，每个视频组件都有唯一的id；设计幻灯片轮播效果，准备好幻灯片需要轮播的图片；

（5）分享功能，需要在Page中定义onShareAppMessage函数，设置该页面的分享信息。

7.2.2　相关知识点

（1）在界面布局的时候，会用到微信小程序的组件，包括view视图容器组件、image图片组件、swiper滑块视图容器组件、scroll-view可滚动视图区域组件、video视频组件等组件的使用；

（2）界面样式设计中，需要使用wxss样式进行界面的美化和渲染；

（3）页签菜单切换的时候，需要获得该页签所对应的id，绑定菜单切换事件；

（4）设计页面分享时，需要使用onShareAppMessage这个API接口，来进行分享；

（5）动态获取糗事列表信息，需要使用wx.request请求获得。

7.3　准备工作

（1）首先需要准备一个AppID，如果没有AppID也没有关系，只不过不能在手机上进行项目的预览，但是在开发工具上开发是没有任何问题的。

（2）在设计列表页时，需要用到一些图标，放置在images/icon文件夹下，如图7.7所示。

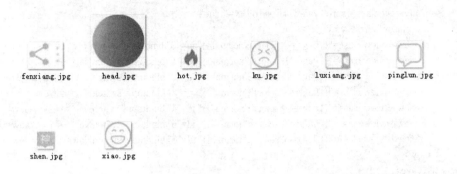

图7.7　图标

7.4　设计流程

我们首先来设计仿糗事百科微信小程序顶部页签菜单左右滑动效果、页签菜单切换效果，切换时页签对应的内容也会随着切换；然后设计糗事列表、视频列表；最后设计分享功能和项目预览。

设计流程

7.4.1　顶部页签菜单滑动效果设计

仿糗事百科微信小程序的顶部页签菜单可以左右滑动，如图7.8所示。

图7.8　顶部页签菜单

（1）新建一个qsbk项目，AppID为wxa7730e0596be9404，把准备好的图片放置在qsbk项目里。

（2）进入到app.json界面里，将窗口背景色设置为黄色（#FFBA1E），标题改为糗事百科，字体颜色设置为白色（white），具体代码如下所示。

```
{
  "pages":[
    "pages/index/index",
    "pages/logs/logs"
  ],
  "window":{
    "backgroundTextStyle":"light",
    "navigationBarBackgroundColor": "#FFBA1E",
    "navigationBarTitleText": "糗事百科",
    "navigationBarTextStyle":"white"
  }
}
```

（3）进入到pages/index/index目录下，清空index.wxml、index.js、index.wxss文件默认生成的内容。

（4）进入到pages/index/index.wxml文件里，设计顶部页签菜单，顶部页签菜单包括3方面内容：页签菜单、"审"字、+，具体代码如下所示。

```
<view class="bg">
  <view class="nav">
    <scroll-view class="scroll-view_H" scroll-x="true" >
    <view class="scroll-view_H">
      <view><view class="{{flag==0?'select':'normal'}}" id="0" bindtap="switchNav">专享</view></view>
      <view><view class="{{flag==1?'select':'normal'}}" id="1" bindtap="switchNav">视频</view></view>
      <view><view class="{{flag==2?'select':'normal'}}" id="2" bindtap="switchNav">糗闻</view></view>
      <view><view class="{{flag==3?'select':'normal'}}" id="3" bindtap="switchNav">纯文</view></view>
      <view><view class="{{flag==4?'select':'normal'}}" id="4" bindtap="switchNav">纯图</view></view>
      <view><view class="{{flag==5?'select':'normal'}}" id="5" bindtap="switchNav">精华</view></view>
      <view><view class="{{flag==6?'select':'normal'}}" id="6" bindtap="switchNav">趣闻</view></view>
    </view>
    </scroll-view>
  </view>
  <view class="opr">
    审
  </view>
  <view class="add">+</view>
</view>
```

（5）进入到pages/index/index.wxss文件里，给顶部菜单内容添加样式，具体代码如下所示。

```
.bg{
  background-color: #FFBA1E;
  height: 50px;
  color: #ffffff;
  display: flex;
  flex-direction: row;
  align-items: center;
}
.nav{
  width: 70%;
  height: 40px;
}
.opr{
  width: 20px;
  height: 20px;
  border-radius:50%;
  font-size: 13px;
  line-height: 20px;
  text-align: center;
  color: #FFBA1E;
  background-color: #ffffff;
  font-weight: bold;
  margin-left: 10px;
}
.add{
  width: 20%;
  height: 50px;
  line-height: 50px;
  text-align: right;
  margin-right:10px;
  font-size: 50px;
}
```

```
.scroll-view_H{
  margin-left: 10px;
  height: 40px;
  display: flex;
  flex-direction: row;
}
.normal{
  width: 40px;
  height: 40px;
  line-height: 40px;
  padding-left:10px;
  padding-right: 10px;
  font-size: 14px;
}
.select{
  width: 40px;
  height: 40px;
  line-height: 40px;
  padding-left:20px;
  padding-right: 20px;
  font-size: 14px;
  font-weight: bold;
}
```

这样就可以实现顶部页签左右滑动效果，如图7.9所示。

图7.9 顶部菜单内容

7.4.2 顶部页签菜单切换效果设计

顶部页签菜单可以进行左右切换，但是还没有实现页签菜单切换效果，单击不同的页签，页签需要呈现为选中状态，同时页签对应的内容也要随着进行切换。

（1）在页签里设计两种样式，一种是select样式，选中时字体加粗；另一种是normal样式，字体不加粗。绑定单击事件switchNav。

（2）进入到pages/index/index.js文件里，定义两个变量：currentTab为当前页签的索引值，flag变量用来控制样式选择，如果flag等于页签对应的id，则呈现为选中状态。使用select这个样式，具体代码如下所示。

```
Page({
  data:{
    currentTab:0,
    flag:0
  }
})
```

（3）添加页签菜单单击绑定事件switchNav，动态地给currentTab和flag变量赋值，具体代码如下所示。

```
Page({
  data:{
    currentTab:0,
```

```
        flag:0
   },
   switchNav:function(e){
     console.log(e);
     var page = this;
     var id = e.target.id;
     if(this.data.currentTab == id){
        return false;
     }else{
        page.setData({currentTab:id});
     }
     page.setData({flag:id});
   }
})
```

这样单击不同页签，页签就会呈现为选中效果，如图7.10所示。

图7.10　页签菜单选中效果

（4）页签菜单进行切换时，对应的内容也随着进行切换，切换设计需要使用swiper滑块视图容器组件。进入到pages/ index/index.wxml文件里，使用swiper进行页签内容布局，具体代码如下所示。

```
<view class="bg">
  <view class="nav">
     <scroll-view class="scroll-view_H" scroll-x="true" >
     <view class="scroll-view_H">
        <view><view class="{{flag==0?'select':'normal'}}" id="0" bindtap="switchNav">专享</view></view>
        <view><view class="{{flag==1?'select':'normal'}}" id="1" bindtap="switchNav">视频</view></view>
        <view><view class="{{flag==2?'select':'normal'}}" id="2" bindtap="switchNav">糗闻</view></view>
        <view><view class="{{flag==3?'select':'normal'}}" id="3" bindtap="switchNav">纯文</view></view>
        <view><view class="{{flag==4?'select':'normal'}}" id="4" bindtap="switchNav">纯图</view></view>
        <view><view class="{{flag==5?'select':'normal'}}" id="5" bindtap="switchNav">精华</view></view>
        <view><view class="{{flag==6?'select':'normal'}}" id="6" bindtap="switchNav">趣闻</view></view>
     </view>
     </scroll-view>
  </view>
  <view class="opr">
     审
  </view>
  <view class="add">+</view>
</view>
<swiper current="{{currentTab}}" style="height:1500px">
  <swiper-item>
     我是专享内容
  </swiper-item>
  <swiper-item>
     我是视频内容
  </swiper-item>
```

```
        <swiper-item>
            我是糗闻内容
        </swiper-item>
        <swiper-item>
            我是纯文内容
        </swiper-item>
        <swiper-item>
            我是纯图内容
        </swiper-item>
        <swiper-item>
            我是精华内容
        </swiper-item>
        <swiper-item>
            我是趣闻内容
        </swiper-item>
</swiper>
```

这样单击页签菜单时，不仅菜单标题呈现为选中状态，而且页签内容也会随着变化，实现了页签菜单和页签内容的联动效果。

7.4.3 糗事列表页设计

糗事列表页面是用来显示糗事内容的页面，每条糗事包括4项内容：发布人头像、发布人昵称、热门图标；发布的糗事内容；糗事好笑数量、评论数量、分享数量；好笑、不好笑、评论、分享的操作按钮图标，如图7.11所示。

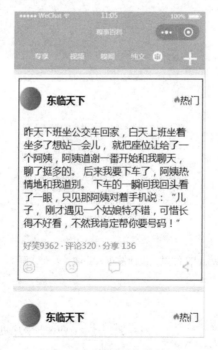

图7.11 糗事列表内容

（1）在pages/index目录下面，新建一个vip.wxml文件，然后进入到pages/index/index.wxml文件里，将vip.wxml文件引入，具体代码如下所示。

```
<view class="bg">
    <view class="nav">
        <scroll-view class="scroll-view_H" scroll-x="true" >
```

```
        <view class="scroll-view_H">
          <view><view class="{{flag==0?'select':'normal'}}" id="0" bindtap="switchNav">专享</view></view>
          <view><view class="{{flag==1?'select':'normal'}}" id="1" bindtap="switchNav">视频</view></view>
          <view><view class="{{flag==2?'select':'normal'}}" id="2" bindtap="switchNav">糗闻</view></view>
          <view><view class="{{flag==3?'select':'normal'}}" id="3" bindtap="switchNav">纯文</view></view>
          <view><view class="{{flag==4?'select':'normal'}}" id="4" bindtap="switchNav">纯图</view></view>
          <view><view class="{{flag==5?'select':'normal'}}" id="5" bindtap="switchNav">精华</view></view>
          <view><view class="{{flag==6?'select':'normal'}}" id="6" bindtap="switchNav">趣闻</view></view>
        </view>
      </scroll-view>
    </view>
    <view class="opr">
      审
    </view>
    <view class="add">+</view>
</view>
<swiper current="{{currentTab}}" style="height:1500px">
    <swiper-item>
        <include src="vip.wxml"/>
    </swiper-item>
    <swiper-item>
        我是视频内容
    </swiper-item>
    <swiper-item>
        我是糗闻内容
    </swiper-item>
    <swiper-item>
        我是纯文内容
    </swiper-item>
    <swiper-item>
        我是纯图内容
    </swiper-item>
    <swiper-item>
        我是精华内容
    </swiper-item>
    <swiper-item>
        我是趣闻内容
    </swiper-item>
</swiper>
```

（2）进入到pages/index/vip.wxml文件里，设计发布人头像、发布人昵称、热门图标，具体代码如下所示。

```
<view class="line"></view>
<view class="item">
  <view class="head">
    <view><image src="../../images/icon/head.jpg" style="width:45px;height:45px;"></image></view>
    <view class="title">东临天下</view>
    <view class="hot"><image src="../../images/icon/hot.jpg" style="width:14px;height:16px;"></image>热门</view>
  </view>
</view>
```

（3）进入到pages/index/index.wxss文件里，给发布人头像、发布人昵称、热门图标添加样式，具体代码如下所示。

```
.bg{
```

```css
    background-color: #FFBA1E;
    height: 50px;
    color: #ffffff;
    display: flex;
    flex-direction: row;
    align-items: center;
}
.nav{
    width: 70%;
    height: 40px;
}
.opr{
    width: 20px;
    height: 20px;
    border-radius:50%;
    font-size: 13px;
    line-height: 20px;
    text-align: center;
    color: #FFBA1E;
    background-color: #ffffff;
    font-weight: bold;
    margin-left: 10px;
}
.add{
    width: 20%;
    height: 50px;
    line-height: 50px;
    text-align: right;
    margin-right:10px;
    font-size: 50px;
}
.scroll-view_H{
    margin-left: 10px;
    height: 40px;
    display: flex;
    flex-direction: row;
}
.normal{
    width: 40px;
    height: 40px;
    line-height: 40px;
    padding-left:10px;
    padding-right: 10px;
    font-size: 14px;
}
.select{
    width: 40px;
    height: 40px;
    line-height: 40px;
    padding-left:20px;
    padding-right: 20px;
    font-size: 14px;
```

```
    font-weight: bold;
}
.line{
    height: 10px;
    background-color: #F2F2F2;
}
.item{
    margin: 10px;
}
.head{
    display: flex;
    flex-direction: row;
    height: 77px;
    align-items: center;
}
.title{
    width: 60%;
    margin-left:10px;
    font-weight: bold;
}
.hot{
    text-align: right;
    width: 30%;
}
```

界面效果如图7.12所示。

图7.12 发布人头像、昵称和热门图标

（4）进入到pages/index/vip.wxml文件里，设计糗事内容、好笑数量、评论数量、分享数量，具体代码如下所示。

```
<view class="line"></view>
<view class="item">
    <view class="head">
        <view><image src="../../images/icon/head.jpg" style="width:45px;height:45px;"></image></view>
        <view class="title">东临天下</view>
        <view class="hot"><image src="../../images/icon/hot.jpg" style="width:14px;height:16px;"></image>热门</view>
    </view>

    <view class="content">
        昨天下班坐公交车回家，白天上班坐着坐多了想站一会儿，\r\n就把座位让给了一个阿姨，阿姨道谢一番开始和我聊天，聊了挺多的。\r\n后来我要下车了，阿姨热情地和我道别。\r\n下车的一瞬间我回头看了一眼，只见那阿姨对着手机说："儿子，\r\n刚才遇见一个姑娘特不错，可惜长得不好看，不然我肯定帮你要号码！"
```

```
        </view>
        <view class="sta">好笑9362 · 评论320 · 分享 136</view>
</view>
```

（5）进入到pages/index/index.wxss文件里，给糗事内容、好笑数量、评论数量、分享数量添加样式，具体代码如下所示。

```css
.bg{
    background-color: #FFBA1E;
    height: 50px;
    color: #ffffff;
    display: flex;
    flex-direction: row;
    align-items: center;
}
.nav{
    width: 70%;
    height: 40px;
}
.opr{
    width: 20px;
    height: 20px;
    border-radius:50%;
    font-size: 13px;
    line-height: 20px;
    text-align: center;
    color: #FFBA1E;
    background-color: #ffffff;
    font-weight: bold;
    margin-left: 10px;
}
.add{
    width: 20%;
    height: 50px;
    line-height: 50px;
    text-align: right;
    margin-right:10px;
    font-size: 50px;
}
.scroll-view_H{
    margin-left: 10px;
    height: 40px;
    display: flex;
    flex-direction: row;
}
.normal{
    width: 40px;
    height: 40px;
    line-height: 40px;
    padding-left:10px;
    padding-right: 10px;
    font-size: 14px;
}
```

```css
.select{
  width: 40px;
  height: 40px;
  line-height: 40px;
  padding-left:20px;
  padding-right: 20px;
  font-size: 14px;
  font-weight: bold;
}
.line{
  height: 10px;
  background-color: #F2F2F2;
}
.item{
  margin: 10px;
}
.head{
  display: flex;
  flex-direction: row;
  height: 77px;
  align-items: center;
}
.title{
  width: 60%;
  margin-left:10px;
  font-weight: bold;
}
.hot{
  text-align: right;
  width: 30%;
}
.content{
  padding: 10px;
  line-height: 25px;
}
.sta{
  padding: 10px;
  color: #999999;
  font-size: 16px;
}
```

（6）进入到pages/index/vip.wxml文件里，设计好笑、不好笑、评论、分享图标，具体代码如下所示。

```
<view class="line"></view>
<view class="item">
  <view class="head">
    <view><image src="../../images/icon/head.jpg" style="width:45px;height:45px;"></image></view>
    <view class="title">东临天下</view>
    <view class="hot"><image src="../../images/icon/hot.jpg" style="width:14px;height:16px;"></image>热门</view>
  </view>

  <view class="content">
    昨天下班坐公交车回家，白天上班坐着坐多了想站一会儿，\r\n就把座位让给了一个阿姨，阿姨道谢一番开始和我聊天，聊了挺多的。\r\n后来我要下车了，阿姨热情地和我道别。\r\n下车的一瞬间我回头看了一眼，只见
```

```
            那阿姨对着手机说："儿子，\r\n刚才遇见一个姑娘特不错，可惜长得不好看，不然我肯定帮你要号码！"
          </view>
          <view class="sta">好笑9362 · 评论320 · 分享136</view>
          <view class="icon">
            <view><image src="../../images/icon/xiao.jpg" style="width:23px;height:23px;"></image></view>
            <view><image src="../../images/icon/ku.jpg" style="width:23px;height:23px;"></image></view>
            <view><image src="../../images/icon/pinglun.jpg" style="width:23px;height:23px;"></image></view>
            <view class="last"><image src="../../images/icon/fenxiang.jpg" style="width:23px;height:23px;"></image></view>
          </view>
        </view>
```

（7）进入到pages/index/index.wxss文件里，给好笑、不好笑、评论、分享图标添加样式，具体代码如下所示。

```
.bg{
    background-color: #FFBA1E;
    height: 50px;
    color: #ffffff;
    display: flex;
    flex-direction: row;
    align-items: center;
}
.nav{
    width: 70%;
    height: 40px;
}
.opr{
    width: 20px;
    height: 20px;
    border-radius:50%;
    font-size: 13px;
    line-height: 20px;
    text-align: center;
    color: #FFBA1E;
    background-color: #ffffff;
    font-weight: bold;
    margin-left: 10px;
}
.add{
    width: 20%;
    height: 50px;
    line-height: 50px;
    text-align: right;
    margin-right:10px;
    font-size: 50px;
}
.scroll-view_H{
    margin-left: 10px;
    height: 40px;
    display: flex;
    flex-direction: row;
}
```

```
.normal{
  width: 40px;
  height: 40px;
  line-height: 40px;
  padding-left:10px;
  padding-right: 10px;
  font-size: 14px;
}
.select{
width: 40px;
  height: 40px;
  line-height: 40px;
  padding-left:20px;
  padding-right: 20px;
  font-size: 14px;
  font-weight: bold;
}
.line{
  height: 10px;
  background-color: #F2F2F2;
}
.item{
  margin: 10px;
}
.head{
  display: flex;
  flex-direction: row;
  height: 77px;
  align-items: center;
}
.title{
  width: 60%;
  margin-left:10px;
  font-weight: bold;
}
.hot{
  text-align: right;
  width: 30%;
}
.content{
  padding: 10px;
  line-height: 25px;
}
.sta{
  padding: 10px;
  color: #999999;
  font-size: 16px;
}
.icon{
  padding: 10px;
  display: flex;
  flex-direction: row;
```

```
}
.icon view{
    margin: 0 auto;
    width: 25%;
}
.last{
    text-align: right;
}
.section{
    text-align: center;
}
```

界面效果如图7.11所示。

（8）把这块区域作为列表的基础单元内容，要使列表循环展现出来，只需要修改发布人头像、发布人昵称、糗事内容以及好笑、评论、分享数量，其余界面布局和样式不需要修改。

7.4.4 视频列表页设计

视频列表页放置一些搞笑视频或者随拍视频，布局时需要使用video视频组件，界面布局方式和糗事列表布局方式一样，如图7.13所示。

图7.13 视频列表

（1）在pages/index目录下面新建一个页面video.wxml，然后将这个页面引入到index.wxml文件里，具体代码如下所示。

```
<view class="bg">
    <view class="nav">
        <scroll-view class="scroll-view_H" scroll-x="true" >
            <view class="scroll-view_H">
                <view><view class="{{flag==0?'select':'normal'}}" id="0" bindtap="switchNav">专享</view></view>
                <view><view class="{{flag==1?'select':'normal'}}" id="1" bindtap="switchNav">视频</view></view>
                <view><view class="{{flag==2?'select':'normal'}}" id="2" bindtap="switchNav">糗闻</view></view>
```

```
              <view><view class="{{flag==3?'select':'normal'}}" id="3" bindtap="switchNav">纯文</view></view>
              <view><view class="{{flag==4?'select':'normal'}}" id="4" bindtap="switchNav">纯图</view></view>
              <view><view class="{{flag==5?'select':'normal'}}" id="5" bindtap="switchNav">精华</view></view>
              <view><view class="{{flag==6?'select':'normal'}}" id="6" bindtap="switchNav">趣闻</view></view>
        </view>
     </scroll-view>
   </view>
   <view class="opr">
        审
   </view>
   <view class="add">+</view>
 </view>
 <swiper current="{{currentTab}}"style="height:1500px">
    <swiper-item>
        <include src="vip.wxml"/>
    </swiper-item>
    <swiper-item>
        <include src="video.wxml"/>
    </swiper-item>
    <swiper-item>
        我是糗闻内容
    </swiper-item>
    <swiper-item>
        我是纯文内容
    </swiper-item>
    <swiper-item>
        我是纯图内容
    </swiper-item>
    <swiper-item>
        我是精华内容
    </swiper-item>
    <swiper-item>
        我是趣闻内容
    </swiper-item>
 </swiper>
```

（2）把vip.wxml文件内容复制到video.wxml文件里，在这个基础上修改video.wxml文件，修改成视频列表页面，具体代码如下所示。

```
<view class="line"></view>
<view class="item">
  <view class="head">
     <view><image src="../../images/icon/head.jpg" style="width:45px;height:45px;"></image></view>
     <view class="title">东临天下</view>
  </view>


     <view class="section">
   <video style="width:320px" id="myVideo"  src="http://wxsnsdy.tc.qq.com/105/20210/snsdyvideodownload?filekey=30280201010421301f0201690402534804102ca905ce620b1241b726bc41dcff44e00204012882540400&bizid=1023&hy=SH&fileparam=302c020101042530230204136ffd93020457e3c4ff02024ef202031e8d7f02030f42400204045a320a0201000400"  controls></video>
</view>
<view class="sta">好笑9362 · 评论320 · 分享 136</view>
  <view class="icon">
```

```
            <view><image src="../../images/icon/xiao.jpg" style="width:23px;height:23px;"></image></view>
            <view><image src="../../images/icon/ku.jpg" style="width:23px;height:23px;"></image></view>
            <view><image src="../../images/icon/pinglun.jpg" style="width:23px;height:23px;"></image></view>
            <view class="last"><image src="../../images/icon/fenxiang.jpg" style="width:23px;height:23px;"></image></view>
    </view>

</view>

<view class="line"></view>
<view class="item">
    <view class="head">
        <view><image src="../../images/icon/head.jpg" style="width:45px;height:45px;"></image></view>
        <view class="title">人生有太多的遗憾</view>
    </view>

        <view class="section">
        <video style="width:320px" id="myVideo1" src="http://wxsnsdy.tc.qq.com/105/20210/snsdyvideodownload?filekey=30280201010421301f0201690402534804102ca905ce620b1241b726bc41dcff44e00204012882540400&bizid=1023&hy=SH&fileparam=302c020101042530230204136ffd93020457e3c4ff02024ef202031e8d7f02030f42400204045a320a0201000400"  controls></video>
</view>
<view class="sta">好笑9362 · 评论320 · 分享 136</view>
    <view class="icon">
            <view><image src="../../images/icon/xiao.jpg" style="width:23px;height:23px;"></image></view>
            <view><image src="../../images/icon/ku.jpg" style="width:23px;height:23px;"></image></view>
            <view><image src="../../images/icon/pinglun.jpg" style="width:23px;height:23px;"></image></view>
            <view class="last"><image src="../../images/icon/fenxiang.jpg" style="width:23px;height:23px;"></image></view>
    </view>

</view>
```

（3）video界面布局样式不需要修改，可以共用index.wxss样式，界面效果如图7.13所示。

7.4.5 分享设计

微信小程序支持分享页功能，可以将当前页分享给好友，单击右上角的 ••• 就可以进行分享，好友看到这个页面时显示的是实时数据，可以直接进入到微信小程序里，如图7.14所示。

图7.14 页面分享

进入到pages/index/index.js文件里，添加onShareAppMessage分享功能接口函数，具体代码如下所示。

```
Page({
  data:{
    currentTab:0,
    flag:0
  },
```

```
    switchNav:function(e){
      console.log(e);
      var page = this;
      var id = e.target.id;
      if(this.data.currentTab == id){
         return false;
      }else{
         page.setData({currentTab:id});
      }
      page.setData({flag:id});
    },
    onShareAppMessage: function () {
      return {
        title: '糗事百科',
        desc: '这里有搞笑的娱乐段子',
        path: '/index/index'
      }
    }
})
```

微信开发者单击分享功能时,界面效果如图7.15所示。

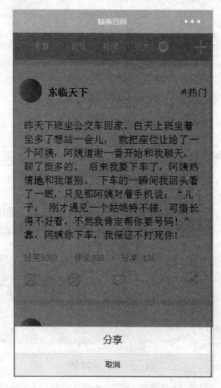

图7.15 分享功能

7.4.6 项目预览

如果项目中有AppID,就可以将项目进行上传,在手机上就可以浏览仿糗事百科微信小程序,如图7.16、图7.17、图7.18所示。

第7章
综合案例：仿糗事百科微信小程序

图7.16　糗事列表　　　　　图7.17　视频列表　　　　　图7.18　分享

7.5　小结

本章主要设计了仿糗事百科微信小程序，重点掌握以下内容：

（1）学会使用scroll-view可滚动视图组件来设计左右滑动效果；

（2）学会设计页签菜单和页签内容联动切换效果；

（3）学会设计糗事列表的布局以及如何使用引入文件复用糗事列表内容，提高开发效率；

（4）学会设计分享功能，借助于微信小程序提供的分享功能接口，可以将当前页面分享给好友；

（5）学会界面布局和给界面添加相关的样式。

第8章
综合案例：仿中国婚博会微信小程序

- 需求描述
- 设计思路及相关知识点
- 准备工作
- 设计流程
- 小结

中国婚博会每个季度举办一次，为结婚人群提供一条龙服务，包括婚纱摄影、酒店、珠宝首饰、婚庆、婚车、婚纱礼服等结婚服务。本章案例模仿制作中国婚博会小程序。

第8章 综合案例：仿中国婚博会微信小程序

参加婚博会需要使用中国婚博会App进行索要门票、领签到礼等环节。由于中国婚博会App软件使用频率不是很高，因此完全可以做一个中国婚博会微信小程序，需要的时候搜索出来使用，它的主要界面如图8.1～图8.4所示。

图8.1 首页

图8.2 全部分类

图8.3 现金券

图8.4 婚博会

8.1 需求描述

仿中国婚博会微信小程序需要完成以下主要功能：

（1）完成底部标签导航设计、首页海报轮播效果设计和宫格导航设计，如图8.5所示。

（2）在首页里，单击全部分类宫格导航的时候，会进入到全部分类导航界面，把婚博会相关内容的导航集成到一个界面里，如图8.6所示。

图8.5 首页设计

图8.6 全部分类界面

（3）在现金券界面里，将各个商户的现金券以列表的形式展现出来，提供全部、默认下拉菜单效果显示，如图8.7所示。

（4）在婚博会界面里，提供索票的界面，填写个人相关信息，可以进行索票，如图8.8所示。

图8.7 现金券界面

图8.8 免费索票

(5) 在填写表单选择获知渠道时，以弹出窗口的形式提供单选列表，供用户选择获知婚博会渠道的情况，如图8.9所示。

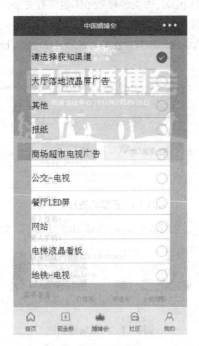

图8.9　获知渠道

8.2　设计思路及相关知识点

8.2.1　设计思路

（1）设计底部标签导航时，准备好底部标签导航的图标，并建立5个相应的页面；设置默认时图片和选中时图片，标签名称采用两种颜色，红色为选中颜色，灰色为默认颜色；

（2）设计幻灯片轮播效果时，准备好幻灯片需要轮播的图片；

（3）设计宫格导航的时候，先把宫格导航的图标和导航名称存放在js后台里，然后动态循环展现出宫格导航；

（4）在设计全部分类导航的时候，有3块区域内容：玩转婚博会、特色分类、我的婚博会。由于这3个区域布局方式一样，可以先设计出一个区域，其余的两个区域直接复制使用即可；

（5）现金券界面设计难点在于下拉菜单筛选条件设计，需要把筛选条件置于页面顶层，在样式里设置z-index:999就可以将其置于最顶层；

（6）婚博会索票界面是常规的表单界面，需要把表单数据提交给后台，保存到本地里。

8.2.2　相关知识点

（1）在界面布局的时候，会用到微信小程序的组件，包括view视图容器组件、image图片组件、swiper滑块视图容器组件、icon图标组件、form表单组件、radio单项选择器组件、checkbox多项选择器组件等组件的使用；

（2）界面样式设计，需要写一些wxss样式进行界面的美化和渲染；

（3）将数据缓存到本地，需要调用wx.setStorageSync这个API接口，进行缓存数据；

（4）界面跳转需要使用wx.navigateTo这个API接口，进行界面跳转。

8.3 准备工作

（1）我们首先需要准备一个AppID，如果没有AppID也没有关系，只不过不能在手机上进行项目的预览，但是在开发工具上开发是没有任何问题的。

（2）底部标签导航，需要有选中图标和默认图标，放置在images/bar文件夹下，如图8.10所示。

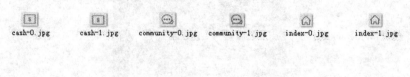

图8.10　底部标签导航图标

（3）需要准备海报轮播的图片，放置在images/haibao文件夹下，如图8.11所示。

图8.11　海报轮播图片

（4）在首页设计宫格导航的时候需要用到一些图标，放置在images/nav文件夹下，如图8.12所示。

图8.12　宫格导航图标

（5）在全部分类界面里，玩转婚博会需要用到的一些图标，放置在images/type/wzhbh文件夹下，如图8.13所示。

图8.13　玩转婚博会图标

（6）在全部分类界面里，特色分类需要用到的一些图标，放置在images/type/tsfl文件夹下，如图8.14所示。

图8.14 特色分类图标

（7）在全部分类界面里，我的婚博会需要用到的一些图标，放置在images/type/wdhbh文件夹下，如图8.15所示。

图8.15 我的婚博会图标

（8）在现金券界面里，需要用到的一些图标，放置在images/cash文件夹下，如图8.16所示。

图8.16 现金券图标

（9）在婚博会索票界面里，需要用到的一些图片，放置在images/marry文件夹下。

8.4 设计流程

我们首先来设计仿中国婚博会微信小程序底部标签导航、海报轮播效果、宫格导航，然后设计全部分类导航界面，再设计现金券下拉菜单筛选条件以及现金券列表页，最后设计婚博会索票界面以及获知渠道弹出层。

设计流程

8.4.1 底部标签导航设计

仿中国婚博会微信小程序，有5个底部标签导航：首页、现金券、婚博会、社区、我的。标签导航选中时导航图标会变为红色图标，导航文字会变为红色文字，如图8.17所示。

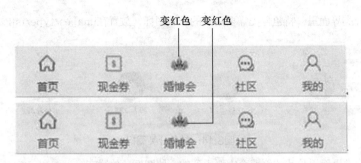

图8.17 底部标签导航选中效果

（1）新建一个hbh项目的微信小程序，将准备好的底部标签导航图标、海报轮播图片、宫格导航图标、现金券图标、婚博会索票图片放置在hbh项目下。

（2）打开app.json配置文件，在pages数组里添加5个页面路径"pages/index/index""pages/cash/cash""pages/marry/marry""pages/community/community""pages/me/me"，保存后会自动生成相应的页面文件夹；删除"pages/logs/logs"页面路径以及对应的文件夹，具体代码如下所示。

```
{
  "pages": [
    "pages/index/index",
    "pages/cash/cash",
    "pages/marry/marry",
    "pages/community/community",
    "pages/me/me"
  ],
  "window": {
    "backgroundTextStyle": "light",
    "navigationBarBackgroundColor": "#fff",
    "navigationBarTitleText": "WeChat",
    "navigationBarTextStyle": "black"
  }
}
```

（3）在window数组里配置窗口导航背景颜色为红色（#D73E3E），导航栏文字为中国婚博会，字体颜色为白色，具体代码如下所示。

```
{
  "pages": [
    "pages/index/index",
    "pages/cash/cash",
    "pages/marry/marry",
    "pages/community/community",
    "pages/me/me"
  ],
  "window": {
    "backgroundTextStyle":"light",
    "navigationBarBackgroundColor": "#D73E3E",
    "navigationBarTitleText": "中国婚博会",
    "navigationBarTextStyle":"white"
  }
}
```

（4）在tabBar对象里配置底部标签导航背景色为灰色（#F3F1EF），文字默认颜色为灰色，选中时为红色（#D73E3E），在list数组里配置底部标签导航对应的页面、导航名称、默认时图标、选中时图标，具体代码如下所示。

```
{
  "pages": [
    "pages/index/index",
    "pages/cash/cash",
    "pages/marry/marry",
    "pages/community/community",
    "pages/me/me"
  ],
  "window": {
    "backgroundTextStyle":"light",
    "navigationBarBackgroundColor": "#D73E3E",
    "navigationBarTitleText": "中国婚博会",
    "navigationBarTextStyle":"white"
  },
  "tabBar": {
    "selectedColor": "#D73E3E",
    "backgroundColor": "#F3F1EF",
    "borderStyle": "white",
    "list": [{
      "pagePath": "pages/index/index",
      "text": "首页",
      "iconPath": "images/bar/index-0.jpg",
      "selectedIconPath": "images/bar/index-1.jpg"
    },{
      "pagePath": "pages/cash/cash",
      "text": "现金券",
      "iconPath": "images/bar/cash-0.jpg",
      "selectedIconPath": "images/bar/cash-1.jpg"
    },{
      "pagePath": "pages/marry/marry",
      "text": "婚博会",
      "iconPath": "images/bar/marry-0.jpg",
      "selectedIconPath": "images/bar/marry-1.jpg"
    },{
      "pagePath": "pages/community/community",
      "text": "社区",
      "iconPath": "images/bar/community-0.jpg",
      "selectedIconPath": "images/bar/community-1.jpg"
    },{
      "pagePath": "pages/me/me",
      "text": "我的",
      "iconPath": "images/bar/me-0.jpg",
      "selectedIconPath": "images/bar/me-1.jpg"
    }]
  }
}
```

这样就完成了仿中国婚博会微信小程序的底部标签导航配置，单击不同的导航，可以切换显示不同的页面，同时导航图标和导航文字会呈现为选中状态。

8.4.2 海报轮播效果设计

海报轮播效果可以在有限的区域内动态地显示不同的广告图片，是很多网站或者App软件都会采用的一种展现方式，在仿中国婚博会微信小程序的首页面里，采用海报轮播效果展示广告图片，如图8.18所示。

图8.18 海报轮播显示

（1）进入到pages/index/index.wxml文件里，采用view、swiper、image进行布局，图片宽度设置为100%，高度设置为176px，具体代码如下所示：

```
<view class="haibao">
    <swiper indicator-dots="{{indicatorDots}}" autoplay="{{autoplay}}" interval="{{interval}}" duration="{{duration}}">
        <block wx:for="{{imgUrls}}">
            <swiper-item>
                <image src="{{item}}" class="silde-image" style="width:100%;height:176px;"></image>
            </swiper-item>
        </block>
    </swiper>
</view>
```

swiper滑块视图容器设置为自动播放(autoplay="true")，自动切换时间间隔为5s(interval="5000")，滑动动画时长为1s(duration="1000")；

采用wx:for循环来显示要展示的图片，从index.js里获取imgUrls图片路径。

（2）进入到pages/ index/index.js文件里，在data对象里定义imgUrls数组，存放海报轮播的图片路径，代码如下：

```
Page({
  data: {
    indicatorDots: false,
    autoplay: true,
    interval: 5000,
    duration: 1000,
    imgUrls: [
      "../../images/haibao/1.jpg",
      "../../images/haibao/2.jpg",
      "../../images/haibao/3.jpg",
      "../../images/haibao/4.jpg"
    ]
  },
```

```
onLoad:function(options){
    // 页面初始化 options为页面跳转所带来的参数
}
})
```

（3）这样就可以实现海报轮播效果，如图8.19、图8.20所示。

图8.19　海报轮播一

图8.20　海报轮播二

8.4.3　宫格导航设计

宫格导航设计是很多App软件都会采用的一种设计方式，我们通过这种宫格导航设计，在首页里就给用户明确的入口，用户可以根据自己的需要进入到相关界面。这就是宫格导航的一个好处，即根据导航名称进入到相应的界面，如图8.21所示。

图8.21　宫格导航设计

（1）进入到pages/index/index.js文件里，首先准备好要显示的图标和导航名称的数据值，然后定义变量navs，将数据值赋值给navs变量，具体代码如下所示：

```
Page({
  data: {
    indicatorDots: false,
    autoplay: true,
    interval: 5000,
    duration: 1000,
    imgUrls: [
      "../../images/haibao/1.jpg",
      "../../images/haibao/2.jpg",
      "../../images/haibao/3.jpg",
      "../../images/haibao/4.jpg"
    ],
    navs: []
  },
  onLoad: function (options) {
    var page = this;
    var navs = this.loadNavData();
    page.setData({ navs: navs });
  },
  loadNavData: function () {
    var navs = [];
    var nav0 = new Object();
    nav0.img = '../../images/nav/dxy.jpg';
    nav0.name = '订喜宴';
    navs[0] = nav0;

    var nav1 = new Object();
    nav1.img = '../../images/nav/phz.jpg';
    nav1.name = '拍婚照';
    navs[1] = nav1;

    var nav2 = new Object();
    nav2.img = '../../images/nav/zhq.jpg';
    nav2.name = '找婚庆';
    navs[2] = nav2;

    var nav3 = new Object();
    nav3.img = '../../images/nav/dhj.jpg';
    nav3.name = '订婚戒';
    navs[3] = nav3;

    var nav4 = new Object();
    nav4.img = '../../images/nav/xhs.jpg';
    nav4.name = '选婚纱';
    navs[4] = nav4;

    var nav5 = new Object();
    nav5.img = '../../images/nav/thp.jpg';
    nav5.name = '淘婚品';
```

```
        navs[5] = nav5;

        var nav6 = new Object();
        nav6.img = '../../images/nav/dmy.jpg';
        nav6.name = '度蜜月';
        navs[6] = nav6;

        var nav7 = new Object();
        nav7.img = '../../images/nav/zhc.jpg';
        nav7.name = '租婚车';
        navs[7] = nav7;

        var nav8 = new Object();
        nav8.img = '../../images/nav/mxn.jpg';
        nav8.name = '美新娘';
        navs[8] = nav8;

        var nav9 = new Object();
        nav9.img = '../../images/nav/qbfl.jpg';
        nav9.name = '全部分类';
        navs[9] = nav9;
        return navs;
    }
})
```

（2）进入到pages/index/index.wxml文件里，进行宫格导航的布局，采用wx:for列表渲染的方式将变量navs值循环显示出来，定义绑定事件navBtn，具体代码如下所示：

```
<view class="haibao">
    <swiper indicator-dots="{{indicatorDots}}" autoplay="{{autoplay}}" interval="{{interval}}" duration="{{duration}}">
        <block wx:for="{{imgUrls}}">
            <swiper-item>
                <image src="{{item}}" class="silde-image" style="width:100%;height:176px;"></image>
            </swiper-item>
        </block>
    </swiper>
</view>
<view class="nav">
<block wx:for="{{navs}}">
    <view class="item" bindtap="navBtn" id="{{index}}">
        <view>
            <image src="{{item.img}}" style="width:58px;height:56px;"></image>
        </view>
        <view>
            {{item.name}}
        </view>
    </view>
</block>
</view>
<view class="hr"></view>
```

（3）进入到pages/index/index.wxss文件里，给宫格导航和间隔线添加相应的样式，具体代码如下所示：

```
.nav{
    text-align: center;
```

```
        }
        .item{
          margin-top:15px;
          text-align: center;
          font-family: "Microsoft YaHei";
          font-size: 13px;
          width: 60px;
          display: inline-block;
          margin-right:10px;
        }
        .hr{
          height: 1px;
          background-color: #cccccc;
          opacity: 0.2;
          margin-top:10px;
        }
```

（4）在app.json里新配置一个全部分类type页面文件路径，再进入到pages/index/index.js文件里，添加navBtn事件用于单击全部分类宫格导航时，跳转到全部分类界面，具体代码如下所示：

```
Page({
  data: {
    indicatorDots: false,
    autoplay: true,
    interval: 5000,
    duration: 1000,
    imgUrls: [
      "../../images/haibao/1.jpg",
      "../../images/haibao/2.jpg",
      "../../images/haibao/3.jpg",
      "../../images/haibao/4.jpg"
    ],
    navs: []
  },
  onLoad: function (options) {
    var page = this;
    var navs = this.loadNavData();
    page.setData({ navs: navs });
  },
  navBtn: function (e) {
    console.log(e);
    var id = e.currentTarget.id;
    if (id == "9") {
      wx.navigateTo({
        url: '../type/type'
      })
    }
  },
  loadNavData: function () {
    var navs = [];
    var nav0 = new Object();
    nav0.img = '../../images/nav/dxy.jpg';
    nav0.name = '订喜宴';
```

```
            navs[0] = nav0;

            var nav1 = new Object();
            nav1.img = '../../images/nav/phz.jpg';
            nav1.name = '拍婚照';
            navs[1] = nav1;

            var nav2 = new Object();
            nav2.img = '../../images/nav/zhq.jpg';
            nav2.name = '找婚庆';
            navs[2] = nav2;

            var nav3 = new Object();
            nav3.img = '../../images/nav/dhj.jpg';
            nav3.name = '订婚戒';
            navs[3] = nav3;

            var nav4 = new Object();
            nav4.img = '../../images/nav/xhs.jpg';
            nav4.name = '选婚纱';
            navs[4] = nav4;

            var nav5 = new Object();
            nav5.img = '../../images/nav/thp.jpg';
            nav5.name = '淘婚品';
            navs[5] = nav5;

            var nav6 = new Object();
            nav6.img = '../../images/nav/dmy.jpg';
            nav6.name = '度蜜月';
            navs[6] = nav6;

            var nav7 = new Object();
            nav7.img = '../../images/nav/zhc.jpg';
            nav7.name = '租婚车';
            navs[7] = nav7;

            var nav8 = new Object();
            nav8.img = '../../images/nav/mxn.jpg';
            nav8.name = '美新娘';
            navs[8] = nav8;

            var nav9 = new Object();
            nav9.img = '../../images/nav/qbfl.jpg';
            nav9.name = '全部分类';
            navs[9] = nav9;
            return navs;
        }
})
```

这样就可以实现宫格导航功能，同时单击全部分类导航时会跳转到全部分类界面，如果想单击其他宫格导航进行跳转，也需要配置相应的跳转界面。

8.4.4 全部分类导航设计

全部分类导航界面集合了中国婚博会所有的导航入口，它可以分为3类导航：玩转婚博会导航，有订喜宴、拍婚照、找婚庆等服务；特色分类导航，有拍写真、美新娘等服务；我的婚博会导航，有现金券、邀请函、签到礼等服务。这3类导航布局方式一样，可以先设计一类导航，然后直接进行复用，如图8.22所示。

图8.22　全部分类界面

（1）进入到pages/index/type.wxml文件里，先来设计玩转婚博会这一类导航，具体代码如下所示：

```html
<view class="content">
<view class="line"></view>
<view class="item">
  <view class="title">玩转婚博会</view>
  <view class="hr"></view>
  <view class="navs">
    <view class="nav">
      <view><image src="../../images/type/wzhbh/dxy.jpg" style="width:38px;height:38px;"></image></view>
      <view>订喜宴</view>
    </view>
    <view class="nav">
      <view><image src="../../images/type/wzhbh/phz.jpg" style="width:38px;height:38px;"></image></view>
      <view>拍婚照</view>
    </view>
    <view class="nav">
      <view><image src="../../images/type/wzhbh/zhq.jpg" style="width:38px;height:38px;"></image></view>
      <view>找婚庆</view>
    </view>
    <view class="nav">
      <view><image src="../../images/type/wzhbh/dhj.jpg" style="width:38px;height:38px;"></image></view>
      <view>订婚戒</view>
    </view>
```

```
        </view>
          <view class="navs">
            <view class="nav">
                <view><image src="../../images/type/wzhbh/xhs.jpg" style="width:38px;height:38px;"></image></view>
                <view>选婚纱</view>
            </view>
            <view class="nav">
                <view><image src="../../images/type/wzhbh/thp.jpg" style="width:38px;height:38px;"></image></view>
                <view>淘婚品</view>
            </view>
            <view class="nav">
                <view><image src="../../images/type/wzhbh/dmy.jpg" style="width:38px;height:38px;"></image></view>
                <view>度蜜月</view>
            </view>
            <view class="nav">
                <view><image src="../../images/type/wzhbh/zhc.jpg" style="width:38px;height:38px;"></image></view>
                <view>租婚车</view>
            </view>
          </view>
        </view>
    </view>
```

（2）进入到pages/index/type.wxss文件里，给玩转婚博会这一类导航添加样式，让它以两行四列的方式展现出来，具体代码如下所示：

```
.content{
    font-family: "Microsoft YaHei";
    background-color: #F0F0F0;
}
.line{
    height: 10px;
}
.item{
    border: 1px solid #cccccc;
    width: 90%;
    margin: 0 auto;
    background-color: #ffffff;
    padding: 10px;
    border-radius: 5px;
}
.hr{
    height: 1px;
    background-color: #cccccc;
    opacity: 0.2;
    margin-top: 10px;
    margin-bottom: 10px;
}
.navs{
    display: flex;
    flex-direction: row;
    text-align: center;
    font-size: 13px;
    margin-bottom:10px;
    padding-top:10px;
}
.nav{
```

```
        margin: 0 auto;
        width:70px;
    }
```
界面效果如图8.23所示。

图8.23　玩转婚博会导航

（3）玩转婚博会导航设计完之后，特色分类导航可以直接复制其内容，然后在这个基础上进行修改，进入到pages/index/type.wxml文件，复制玩转婚博会区域，修改成特色分类导航内容，具体代码如下所示：

```
<view class="content">
<view class="line"></view>
<view class="item">
    <view class="title">玩转婚博会</view>
    <view class="hr"></view>
    <view class="navs">
        <view class="nav">
            <view><image src="../../images/type/wzhbh/dxy.jpg" style="width:38px;height:38px;"></image></view>
            <view>订喜宴</view>
        </view>
        <view class="nav">
            <view><image src="../../images/type/wzhbh/phz.jpg" style="width:38px;height:38px;"></image></view>
            <view>拍婚照</view>
        </view>
        <view class="nav">
            <view><image src="../../images/type/wzhbh/zhq.jpg" style="width:38px;height:38px;"></image></view>
            <view>找婚庆</view>
        </view>
        <view class="nav">
            <view><image src="../../images/type/wzhbh/dhj.jpg" style="width:38px;height:38px;"></image></view>
            <view>订婚戒</view>
        </view>
    </view>
    <view class="navs">
        <view class="nav">
            <view><image src="../../images/type/wzhbh/xhs.jpg" style="width:38px;height:38px;"></image></view>
            <view>选婚纱</view>
        </view>
        <view class="nav">
            <view><image src="../../images/type/wzhbh/thp.jpg" style="width:38px;height:38px;"></image></view>
            <view>淘婚品</view>
        </view>
        <view class="nav">
```

```
            <view><image src="../../images/type/wzhbh/dmy.jpg" style="width:38px;height:38px;"></image></view>
            <view>度蜜月</view>
        </view>
        <view class="nav">
            <view><image src="../../images/type/wzhbh/zhc.jpg" style="width:38px;height:38px;"></image></view>
            <view>租婚车</view>
        </view>
    </view>
</view>

<view class="line"></view>
<view class="item">
    <view class="title">特色分类</view>
    <view class="hr"></view>
    <view class="navs">
        <view class="nav">
            <view><image src="../../images/type/tsfl/pxz.jpg" style="width:38px;height:38px;"></image></view>
            <view>拍写真</view>
        </view>
        <view class="nav">
            <view><image src="../../images/type/tsfl/mxn.jpg" style="width:38px;height:38px;"></image></view>
            <view>美新娘</view>
        </view>
        <view class="nav">
            <view><image src="../../images/type/tsfl/zxj.jpg" style="width:38px;height:38px;"></image></view>
            <view>装新家</view>
        </view>
        <view class="nav">
            <view><image src="../../images/type/tsfl/yyt.jpg" style="width:38px;height:38px;"></image></view>
            <view>孕婴童</view>
        </view>
    </view>
</view>
</view>
```

界面效果如图8.24所示。

图8.24 特色分类导航

（4）我的婚博会导航区域也用此方法进行设计，进入到pages/index/type.wxml文件里，复制玩转婚博会区域，修改成我的婚博会导航内容，具体代码如下所示：

```
<view class="content">
<view class="line"></view>
<view class="item">
  <view class="title">玩转婚博会</view>
  <view class="hr"></view>
  <view class="navs">
    <view class="nav">
      <view><image src="../../images/type/wzhbh/dxy.jpg" style="width:38px;height:38px;"></image></view>
      <view>订喜宴</view>
    </view>
    <view class="nav">
      <view><image src="../../images/type/wzhbh/phz.jpg" style="width:38px;height:38px;"></image></view>
      <view>拍婚照</view>
    </view>
    <view class="nav">
      <view><image src="../../images/type/wzhbh/zhq.jpg" style="width:38px;height:38px;"></image></view>
      <view>找婚庆</view>
    </view>
    <view class="nav">
      <view><image src="../../images/type/wzhbh/dhj.jpg" style="width:38px;height:38px;"></image></view>
      <view>订婚戒</view>
    </view>
  </view>
   <view class="navs">
    <view class="nav">
      <view><image src="../../images/type/wzhbh/xhs.jpg" style="width:38px;height:38px;"></image></view>
      <view>选婚纱</view>
    </view>
    <view class="nav">
      <view><image src="../../images/type/wzhbh/thp.jpg" style="width:38px;height:38px;"></image></view>
      <view>淘婚品</view>
    </view>
    <view class="nav">
      <view><image src="../../images/type/wzhbh/dmy.jpg" style="width:38px;height:38px;"></image></view>
      <view>度蜜月</view>
    </view>
    <view class="nav">
      <view><image src="../../images/type/wzhbh/zhc.jpg" style="width:38px;height:38px;"></image></view>
      <view>租婚车</view>
    </view>
  </view>
</view>

<view class="line"></view>
<view class="item">
  <view class="title">特色分类</view>
  <view class="hr"></view>
  <view class="navs">
    <view class="nav">
```

```
                <view><image src="../../images/type/tsfl/pxz.jpg" style="width:38px;height:38px;"></image></view>
                <view>拍写真</view>
            </view>
            <view class="nav">
                <view><image src="../../images/type/tsfl/mxn.jpg" style="width:38px;height:38px;"></image></view>
                <view>美新娘</view>
            </view>
            <view class="nav">
                <view><image src="../../images/type/tsfl/zxj.jpg" style="width:38px;height:38px;"></image></view>
                <view>装新家</view>
            </view>
            <view class="nav">
                <view><image src="../../images/type/tsfl/yyt.jpg" style="width:38px;height:38px;"></image></view>
                <view>孕婴童</view>
            </view>
        </view>
    </view>

    <view class="line"></view>
    <view class="item">
        <view class="title">我的婚博会</view>
        <view class="hr"></view>
        <view class="navs">
            <view class="nav">
                <view><image src="../../images/type/wdhbh/xjq.jpg" style="width:38px;height:38px;"></image></view>
                <view>现金券</view>
            </view>
            <view class="nav">
                <view><image src="../../images/type/wdhbh/yqh.jpg" style="width:38px;height:38px;"></image></view>
                <view>邀请函</view>
            </view>
            <view class="nav">
                <view><image src="../../images/type/wdhbh/qdl.jpg" style="width:38px;height:38px;"></image></view>
                <view>签到礼</view>
            </view>
            <view class="nav">
                <view><image src="../../images/type/wdhbh/dhl.jpg" style="width:38px;height:38px;"></image></view>
                <view>兑好礼</view>
            </view>
        </view>
        <view class="navs">
            <view class="nav">
                <view><image src="../../images/type/wdhbh/wdsq.jpg" style="width:38px;height:38px;"></image></view>
                <view>我的社区</view>
            </view>
            <view class="nav">
                <view><image src="../../images/type/wdhbh/yy.jpg" style="width:38px;height:38px;"></image></view>
                <view>预约到婚博会</view>
            </view>
            <view class="nav">
                <view><image src="../../images/type/wdhbh/tj.jpg" style="width:38px;height:38px;"></image></view>
                <view>推荐好友送现金</view>
```

```
            </view>
            <view class="nav">
            </view>
        </view>
    </view>
    <view class="line"></view>

</view>
```

界面效果如图8.25所示。

图8.25 我的婚博会导航

（5）进入到pages/type/type.json文件里，修改窗口标题，具体代码如下所示：

```
{
    "navigationBarTitleText": "全部分类"
}
```

微信小程序在开发过程中总会遇到相同或者类似布局的区域，这时就要新设计一个区域，如果能将其提炼成模板，可以提炼成模板使用；不能的话，可以直接复制这个区域，在这个基础上进行修改，从而可以提高开发效率。

8.4.5 现金券下拉菜单筛选条件设计

很多App都会采用下拉菜单作为筛选条件，下拉菜单里会展现出很多筛选条件供用户选择，现金券界面也是采用下拉菜单的方式来进行条件筛选的，它分为两类条件：全部、默认。如图8.26、图8.27所示。

（1）进入到pages/index/cash.wxml文件里，使用dl、dt、dd进行筛选条件的布局，绑定菜单切换单击事件tapMainMenu，具体代码如下所示。

第8章 综合案例：仿中国婚博会微信小程序

图8.26 全部筛选条件

图8.27 默认筛选条件

```
<dl class="menu">
    <dt data-index="0" bindtap="tapMainMenu" class="{{currentTab==0?'select':'defalut'}}">全部</dt>
    <dd class="{{subMenuDispaly[0]}}">
        <ul>
            <li class="select">全部</li>
            <li>订喜宴</li>
            <li>拍婚照</li>
            <li>找婚庆</li>
            <li>订婚戒</li>
            <li>选婚纱</li>
            <li>淘婚品</li>
            <li>度蜜月</li>
            <li>美新娘</li>
            <li>拍写真</li>
        </ul>
    </dd>
    <dt data-index="1" bindtap="tapMainMenu" class="{{currentTab==1?'select':'defalut'}}">默认</dt>
    <dd class="{{subMenuDispaly[1]}}">
        <ul>
            <li class="select">默认</li>
            <li>最新</li>
            <li>最热</li>
        </ul>
    </dd>
</dl>
```

（2）进入到pages/index/cash.wxss文件里，给下拉菜单筛选条件添加样式，具体代码如下所示。

```
.menu{
    display: block;
    height: 38px;
}
.menu dt{
```

```css
    font-size: 13px;
    float: left;
    width: 49.7%;
    height: 38px;
    border-right: 1px solid #f2f2f2;
    border-bottom: 1px solid #f2f2f2;
    text-align: center;
    line-height: 38px;
    background-color: #ffffff;
}
.menu dd{
    position: absolute;
    width: 100%;
    top:40px;
    left: 0;
    z-index: 999;
}
.menu li{
    font-size: 14px;
    background-color: #F3F1EF;
    line-height: 40px;
    display: block;
    padding-left: 8px;
    border-bottom: 1px solid #ffffff;
}
.select{
    color:red;
}
.show{
    display: block;
}
.hidden{
    display: none;
}
```

界面效果如图8.28所示。

图8.28　筛选条件

（3）进入到pages/index/cash.js文件里，定义两个变量：subMenuDispaly用来控制两个筛选条件内容的显示与隐藏，currentTab用来控制筛选菜单选中效果，并添加菜单切换事件tapMainMenu，具体代码如下所示。

```
function initSubMenuDisplay(){
  return ['hidden','hidden'];
}
Page({
  data:{
    subMenuDispaly:initSubMenuDisplay(),
    currentTab:-1
  },
  tapMainMenu:function(e){
    console.log(e);
    var index = parseInt(e.currentTarget.dataset.index);
    console.log(index);
    var newSubMenuDisplay = initSubMenuDisplay();
    if(this.data.subMenuDispaly[index] == 'hidden'){
      newSubMenuDisplay[index] = 'show';
      this.setData({currentTab:index});
    }else{
      newSubMenuDisplay[index] = 'hidden';
      this.setData({currentTab:-1});
    }
    this.setData({subMenuDispaly:newSubMenuDisplay});
  }
})
```

这样就可以实现下拉菜单筛选条件设计，单击筛选条件菜单名称，菜单名称会呈现为红色选中效果，同时显示出下拉筛选条件选项；单击另一个筛选条件菜单时，会切换显示，以达到一种动态效果。

8.4.6 现金券列表页设计

现金券列表页用来展示商家的现金优惠券，列表里包括商家的图片、商家名称、现金优惠额以及申请情况，如图8.29所示。

图8.29 现金优惠券列表

（1）进入到pages/index/cash.wxml文件里，先来设计一条商家的现金优惠券信息，包括商家图片、商家名称、现金优惠额、申请情况，具体代码如下所示。

```
<dl class="menu">
    <dt data-index="0" bindtap="tapMainMenu" class="{{currentTab==0?'select':'defalut'}}">全部</dt>
    <dd class="{{subMenuDispaly[0]}}">
        <ul>
            <li class="select">全部</li>
            <li>订喜宴</li>
            <li>拍婚照</li>
            <li>找婚庆</li>
            <li>订婚戒</li>
            <li>选婚纱</li>
            <li>淘婚品</li>
            <li>度蜜月</li>
            <li>美新娘</li>
            <li>拍写真</li>
        </ul>
    </dd>
    <dt data-index="1" bindtap="tapMainMenu" class="{{currentTab==1?'select':'defalut'}}">默认</dt>
    <dd class="{{subMenuDispaly[1]}}">
        <ul>
            <li class="select">默认</li>
            <li>最新</li>
            <li>最热</li>
        </ul>
    </dd>
</dl>
 <view class="items">
    <view class="item">
        <view><image src="../../images/cash/sh1.jpg" style=" width:93px;height:73px;" ></image></view>
        <view class="des">
            <view class="title">
                施华洛婚纱摄影<text class="Apply">已有4人申请</text>
            </view>
            <view class="hr"></view>
            <view class="price">￥200-300</view>
        </view>
    </view>
    <view class="line"></view>
</view>
```

（2）进入到pages/index/cash.wxss文件里，给现金优惠券信息添加样式，将其分为左右两列，左侧是商家图片，右侧是现金优惠券相关信息，具体代码如下所示。

```
.menu{
    display: block;
    height: 38px;
}
.menu dt{
    font-size: 13px;
    float: left;
    width: 49.7%;
    height: 38px;
```

```css
    border-right: 1px solid #f2f2f2;
    border-bottom: 1px solid #f2f2f2;
    text-align: center;
    line-height: 38px;
    background-color: #ffffff;
}
.menu dd{
    position: absolute;
    width: 100%;
    top:40px;
    left: 0;
    z-index: 999;
}
.menu li{
    font-size: 14px;
    background-color: #F3F1EF;
    line-height: 40px;
    display: block;
    padding-left: 8px;
    border-bottom: 1px solid #ffffff;
}
.select{
     color:red;
}
.show{
    display: block;
}
.hidden{
    display: none;
}
.item{
    margin: 10px;
    display: flex;
    flex-direction: row;
}
.des{
    margin-left: 10px;
    width: 100%;
}
.Apply{
    font-size: 12px;
    color: #cccccc;
    position: absolute;
    right: 10px;
    margin-top:5px;
}
.hr{
    height: 1px;
    background-color: #f2f2f2;
    margin-top:5px;
    margin-bottom: 5px;
}
```

```css
.price{
    font-size: 25px;
    color: red;
    font-weight: bold;
}
.line{
    height: 1px;
    background-color: #f2f2f2;
    margin-top:10px;
    margin-bottom: 10px;
}
```

界面效果如图8.30所示。

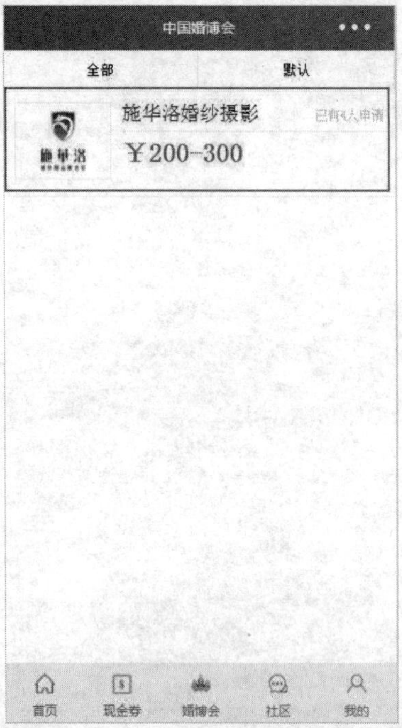

图8.30 现金优惠券布局

(3) 进入到pages/index/cash.wxml文件里,复制设计好的现金优惠券信息,在这个基础上进行修改,具体代码如下所示。

```
<dl class="menu">
    <dt data-index="0" bindtap="tapMainMenu" class="{{currentTab==0?'select':'defalut'}}">全部</dt>
    <dd class="{{subMenuDispaly[0]}}">
        <ul>
            <li class="select">全部</li>
            <li>订喜宴</li>
            <li>拍婚照</li>
            <li>找婚庆</li>
            <li>订婚戒</li>
            <li>选婚纱</li>
            <li>淘婚品</li>
            <li>度蜜月</li>
            <li>美新娘</li>
```

```
                    <li>拍写真</li>
                </ul>
            </dd>
    <dt data-index="1" bindtap="tapMainMenu" class="{{currentTab==1?'select':'defalut'}}">默认</dt>
    <dd class="{{subMenuDispaly[1]}}">
        <ul>
            <li class="select">默认</li>
            <li>最新</li>
            <li>最热</li>
        </ul>
    </dd>
</dl>
<view class="items">
    <view class="item">
        <view><image src="../../images/cash/shl.jpg" style=" width:93px;height:73px;"></image></view>
        <view class="des">
            <view class="title">
                施华洛婚纱摄影<text class="Apply">已有4人申请</text>
            </view>
            <view class="hr"></view>
            <view class="price">￥200-300</view>
        </view>
    </view>
    <view class="line"></view>

    <view class="item">
        <view><image src="../../images/cash/fsl.jpg" style=" width:93px;height:73px;"></image></view>
        <view class="des">
            <view class="title">
                枫树林高端婚礼<text class="Apply">已有14人申请</text>
            </view>
            <view class="hr"></view>
            <view class="price">￥100</view>
        </view>
    </view>
    <view class="line"></view>

    <view class="item">
        <view><image src="../../images/cash/hjls.jpg" style=" width:93px;height:73px;"></image></view>
        <view class="des">
            <view class="title">
                花嫁丽舍一站式<text class="Apply">已有129人申请</text>
            </view>
            <view class="hr"></view>
            <view class="price">￥1000</view>
        </view>
    </view>
    <view class="line"></view>

    <view class="item">
        <view><image src="../../images/cash/sese.jpg" style=" width:93px;height:73px;"></image></view>
        <view class="des">
```

```
            <view class="title">
                SeSe婚礼国王<text class="Apply">已有117人申请</text>
            </view>
            <view class="hr"></view>
            <view class="price">￥100</view>
        </view>
    </view>
    <view class="line"></view>

    <view class="item">
        <view><image src="../../images/cash/wly.jpg" style=" width:93px;height:73px;" ></image></view>
        <view class="des">
            <view class="title">
                婚宴定制酒<text class="Apply">已有4人申请</text>
            </view>
            <view class="hr"></view>
            <view class="price">￥100</view>
        </view>
    </view>
    <view class="line"></view>

    <view class="item">
        <view><image src="../../images/cash/spl.jpg" style=" width:93px;height:73px;" ></image></view>
        <view class="des">
            <view class="title">
                诗普琳珠宝<text class="Apply">已有45人申请</text>
            </view>
            <view class="hr"></view>
            <view class="price">￥100</view>
        </view>
    </view>
    <view class="line"></view>

    <view class="item">
        <view><image src="../../images/cash/cym.jpg" style=" width:93px;height:73px;" ></image></view>
        <view class="des">
            <view class="title">
                朝阳门私属婚礼空间<text class="Apply">已有3人申请</text>
            </view>
            <view class="hr"></view>
            <view class="price">￥800</view>
        </view>
    </view>
    <view class="line"></view>
</view>
```

界面效果如图8.31所示。

（4）进入到pages/cash/cash.json文件里，修改窗口标题，具体代码如下所示：

```
{
    "navigationBarTitleText": "现金券"
}
```

图8.31 现金券界面

8.4.7 婚博会索票界面设计

婚博会索票界面用来索要中国婚博会门票，门票是参加婚博会的入门凭证，而索要门票需要填写自己和爱人以及婚礼的相关信息，如图8.32所示。

图8.32 索票表单

（1）进入到pages/marry/marry.wxml文件里，设计婚博会宣传海报和索票按钮的布局，具体代码如下所示：

```
<view class="content">
  <view>
    <image src="../../images/marry/xuanchuan.jpg" style="width:100%;height:220px;"></image>
  </view>
  <view class="ticket">
     <view class="first"><icon type="success"></icon>我第一次索票</view>
     <view class="second"><icon type="success"></icon>我以前索过票</view>
  </view>
</view>
```

（2）进入到pages/marry/marry.wxss文件里，给索票按钮添加样式，具体代码如下所示：

```
.ticket{
    display: flex;
    flex-direction: row;
    width: 100%;
    text-align: center;
    margin-bottom: 30px;
}
.first{
    border:1px solid red;
    width: 45%;
     margin: 0 auto;
    background-color: #D60210;
    height: 35px;
    line-height: 35px;
    font-size: 14px;
     color: #ffffff;
}
.second{
    border:1px solid red;
    width: 45%;
    margin: 0 auto;
    background-color: #FDA6AE;
    height: 35px;
    line-height: 35px;
    font-size: 14px;
}
```

界面效果如图8.33所示。

图8.33　索票按钮

（3）进入到pages/marry/marry.wxml文件里，设计索票需要填写的表单，需要用到input文本框组件、checkbox多项选择器组件、button按钮组件、radio单项选择器组件、form组件，具体代码如下所示：

```
<view class="content">
    <view>
        <image src="../../images/marry/xuanchuan.jpg" style="width:100%;height:220px;"></image>
    </view>
    <view class="ticket">
        <view class="first"><icon type="success"></icon>我第一次索票</view>
        <view class="second"><icon type="success"></icon>我以前索过票</view>
    </view>

<form bindsubmit="formSubmit">
    <view class="item">
        <view class="name">姓名：</view>
        <view class="val"><input name="name" type="text"/></view>
    </view>
    <view class="item">
        <view class="name">手机号：</view>
        <view class="val"><input name="mobile" type="text"/></view>
    </view>
    <view class="item">
        <view class="name">爱人姓名：</view>
        <view class="val"><input name="lovename" type="text"/></view>
    </view>
    <view class="item">
        <view class="name">爱人手机：</view>
        <view class="val"><input name="lovemobile" type="text"/></view>
    </view>
    <view class="item">
        <view class="name">快递地址：</view>
        <view class="val"><input name="address" type="text" placeholder="情侣套票免费快递到家" placeholder-class="holder"/></view>
    </view>
    <view class="item">
        <view class="name">婚期：</view>
        <view class="val"><input name="date" type="text" placeholder="请输入日期" placeholder-class="holder"/></view>
    </view>
    <view class="item">
        <view class="name">需要筹备：</view>
        <view class="box">
            <checkbox-group name="box">
                <checkbox value="订喜宴">订喜宴</checkbox>
                <checkbox value="找婚庆">找婚庆</checkbox>
                <checkbox value="拍婚照">拍婚照</checkbox>
                <checkbox value="订婚戒">订婚戒</checkbox>
                <checkbox value="选婚纱">选婚纱</checkbox>
                <checkbox value="度蜜月">度蜜月</checkbox>
                <checkbox value="淘婚品">淘婚品</checkbox>
                <checkbox value="美新娘">美新娘</checkbox>
                <checkbox value="租婚车">租婚车</checkbox>
            </checkbox-group>
        </view>
    </view>
```

```
        <view class="item">
            <view class="name">获知渠道: </view>
            <view><button class="way" bindtap="selectWay">{{way}}</button></view>
        </view>
        <button class="btn" form-type="submit">免费索票</button>
    </form>
</view>
```

（4）进入到pages/marry/marry.wxss文件里，给表单添加样式，具体代码如下所示：

```
.ticket{
    display: flex;
    flex-direction: row;
    width: 100%;
    text-align: center;
    margin-bottom: 30px;
}
.first{
    border:1px solid red;
    width: 45%;
    margin: 0 auto;
    background-color: #D60210;
    height: 35px;
    line-height: 35px;
    font-size: 14px;
    color: #ffffff;
}
.second{
    border:1px solid red;
    width: 45%;
    margin: 0 auto;
    background-color: #FDA6AE;
    height: 35px;
    line-height: 35px;
    font-size: 14px;
}
.item{
    margin: 10px;
    display: flex;
    flex-direction: row;
}
.name{
    width: 100px;
    text-align: right;
    height: 28px;
    line-height: 28px;
    font-size: 15px;
}
.val{
    width: 230px;
    height: 28px;
    border:1px solid #cccccc;
    border-radius: 5px;
    font-size:13px;
}
.holder{
```

```css
        font-size: 13px;
        margin-left: 5px;
        color: #999999;
}
.box{
        font-size: 13px;
        width: 300px;
        margin-left: 10px;
}
.box checkbox{
        margin-right: 10px;
        margin-top: 10px;
}
.btn{
        width:220px;
        height: 45px;
        line-height: 45px;
        background-color: #D50310;
        color: #ffffff;
        border-radius: 50px;
}
.way{
        width: 230px;
        height: 30px;
        font-size: 13px;
        text-align: left;
}
```

界面效果如图8.34所示。

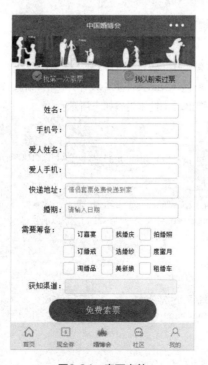

图8.34　索票表单

使用form组件绑定bindsubmit="formSubmit"事件，表单项添加name属性，然后单击button按钮绑定

form-type="submit"属性就可以提交表单。现在获知渠道还没有内容，单击获知渠道会以弹出层的形式展现出来。

8.4.8 获知渠道弹出层设计

获知渠道是以弹出层radio单项选择器组件的列表展现出来的，而且只能选择一种获知渠道，如图8.35所示。

图8.35 获知渠道界面

（1）进入到pages/marry/marry.wxml文件里，设计弹出框布局，并使用radio单项选择器进行列表布局，通过flag变量来控制是否显示弹出层，具体代码如下所示：

```
<view class="content">
  <view>
    <image src="../../images/marry/xuanchuan.jpg" style="width:100%;height:220px;"></image>
  </view>
  <view class="ticket">
    <view class="first"><icon type="success"></icon>我第一次索票</view>
    <view class="second"><icon type="success"></icon>我以前索过票</view>
  </view>

  <form bindsubmit="formSubmit">
    <view class="item">
      <view class="name">姓名：</view>
      <view class="val"><input name="name" type="text"/></view>
    </view>
    <view class="item">
      <view class="name">手机号：</view>
      <view class="val"><input name="mobile" type="text"/></view>
    </view>
```

```
        <view class="item">
            <view class="name">爱人姓名：</view>
            <view class="val"><input name="lovename" type="text"/></view>
        </view>
        <view class="item">
            <view class="name">爱人手机：</view>
            <view class="val"><input name="lovemobile" type="text"/></view>
        </view>
        <view class="item">
            <view class="name">快递地址：</view>
            <view class="val"><input name="address" type="text" placeholder="情侣套票免费快递到家" placeholder-class="holder"/></view>
        </view>
        <view class="item">
            <view class="name">婚期：</view>
            <view class="val"><input name="date" type="text" placeholder="请输入日期" placeholder-class="holder"/></view>
        </view>
        <view class="item">
            <view class="name">需要筹备：</view>
            <view class="box">
                <checkbox-group name="box">
                    <checkbox value="订喜宴">订喜宴</checkbox>
                    <checkbox value="找婚庆">找婚庆</checkbox>
                    <checkbox value="拍婚照">拍婚照</checkbox>
                    <checkbox value="订婚戒">订婚戒</checkbox>
                    <checkbox value="选婚纱">选婚纱</checkbox>
                    <checkbox value="度蜜月">度蜜月</checkbox>
                    <checkbox value="淘婚品">淘婚品</checkbox>
                    <checkbox value="美新娘">美新娘</checkbox>
                    <checkbox value="租婚车">租婚车</checkbox>
                </checkbox-group>
            </view>
        </view>
         <view class="item">
            <view class="name">获知渠道：</view>
            <view><button class="way" bindtap="selectWay">{{way}}</button></view>
        </view>
        <button class="btn" form-type="submit">免费索票</button>
    </form>
</view>
<view class="{{flag=='0'?'bg':'hideBg'}}">
    <view class="radioBg">
        <radio-group bindchange="radioChange">
            <view class="radioItem">
                <view class="radioName">请选择获知渠道</view>
                <view class="radioVal"><radio value="请选择获知渠道" checked/></view>
            </view>
            <view class="radioItem">
                <view class="radioName">大厅落地液晶屏广告</view>
                <view class="radioVal"><radio value="大厅落地液晶屏广告"/></view>
            </view>
```

```
            <view class="radioItem">
                <view class="radioName">其他</view>
                <view class="radioVal"><radio value="其他"/></view>
            </view>
            <view class="radioItem">
                <view class="radioName">报纸</view>
                <view class="radioVal"><radio value="报纸"/></view>
            </view>
            <view class="radioItem">
                <view class="radioName">商场超市电视广告</view>
                <view class="radioVal"><radio value="商场超市电视广告"/></view>
            </view>
            <view class="radioItem">
                <view class="radioName">公交-电视</view>
                <view class="radioVal"><radio value="公交-电视"/></view>
            </view>
            <view class="radioItem">
                <view class="radioName">餐厅LED屏</view>
                <view class="radioVal"><radio value="餐厅LED屏"/></view>
            </view>
            <view class="radioItem">
                <view class="radioName">网站</view>
                <view class="radioVal"><radio value="网站"/></view>
            </view>
            <view class="radioItem">
                <view class="radioName">电梯液晶看板</view>
                <view class="radioVal"><radio value="电梯液晶看板"/></view>
            </view>
            <view class="radioItem">
                <view class="radioName">地铁-电视</view>
                <view class="radioVal"><radio value="地铁-电视"/></view>
            </view>
        </radio-group>
    </view>
</view>
```

（2）进入到pages/marry/marry.wxss文件里，给弹出框添加样式，置于页面顶层需要使用z-index属性，具体代码如下所示：

```
.ticket{
    display: flex;
    flex-direction: row;
    width: 100%;
    text-align: center;
    margin-bottom: 30px;
}
.first{
    border:1px solid red;
    width: 45%;
    margin: 0 auto;
    background-color: #D60210;
    height: 35px;
    line-height: 35px;
```

```css
        font-size: 14px;
        color: #ffffff;
}
.second{
        border:1px solid red;
        width: 45%;
        margin: 0 auto;
        background-color: #FDA6AE;
        height: 35px;
        line-height: 35px;
        font-size: 14px;
}
.item{
        margin: 10px;
        display: flex;
        flex-direction: row;
}
.name{
        width: 100px;
        text-align: right;
        height: 28px;
        line-height: 28px;
        font-size: 15px;
}
.val{
        width: 230px;
        height: 28px;
        border:1px solid #cccccc;
        border-radius: 5px;
        font-size:13px;
}
.holder{
        font-size: 13px;
        margin-left: 5px;
        color: #999999;
}
.box{
        font-size: 13px;
        width: 300px;
        margin-left: 10px;
}
.box checkbox{
        margin-right: 10px;
        margin-top: 10px;
}
.btn{
        width:220px;
        height: 45px;
        line-height: 45px;
        background-color: #D50310;
        color: #ffffff;
        border-radius: 50px;
```

```
        }
        .way{
            width: 230px;
            height: 30px;
            font-size: 13px;
            text-align: left;
        }
        .bg{
             display: block;
            background-color: #cccccc;
            width: 100%;
            height: 750px;
            position: absolute;
            top:0px;
            left:0px;
            z-index: 999;
            opacity: 0.9;
        }
        .radioBg{
            background-color: #ffffff;
            width: 80%;
            height: 500px;
            margin: 0 auto;
            margin-top:20px;
        }
        .radioItem{
            display: flex;
            flex-direction: row;
            height: 50px;
            align-items: center;
            border-bottom: 1px solid #cccccc;
        }
        .radioName{
            width: 90%;
        }
        .hideBg{
            display: none;
        }
```

（3）进入到pages/marry/marry.js文件里，定义两个变量：flag等于0代表显示弹出框，flag等于1代表不显示弹出框；变量way是获知渠道按钮的名称，添加selectWay绑定事件，单击按钮时，显示弹出框，具体代码如下所示：

```
Page({
  data:{
    flag:'1',
    way:'请选择获知渠道'
  },
  selectWay:function(){
    this.setData({flag:'0'});
  }
})
```

（4）单击获知渠道时，弹出框显示获知渠道列表，界面效果如图8.36所示。

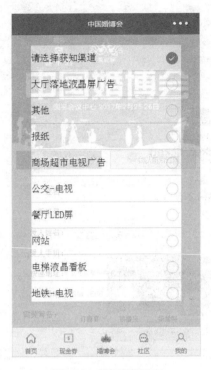

图8.36 获知渠道列表

（5）进入到pages/marry/marry.js文件里，给单项选择器添加radioChange绑定事件，同时设置获知渠道按钮的名称，具体代码如下所示。

```
Page({
  data:{
    flag:'1',
    way:'请选择获知渠道'
  },
  selectWay:function(){
    this.setData({flag:'0'});
  },
  radioChange:function(e){
    console.log(e);
    var way = e.detail.value;
    this.setData({flag:'1'});
    this.setData({way:way});
  }
})
```

（6）进入到pages/marry/marry.js文件里，添加form表单绑定formSubmit事件，可以获取表单提交的内容，将表单提交的内容保存到本地，具体代码如下所示。

```
Page({
  data:{
    flag:'1',
    way:'请选择获知渠道'
  },
  selectWay:function(){
    this.setData({flag:'0'});
  },
  radioChange:function(e){
```

```
      console.log(e);
      var way = e.detail.value;
      this.setData({flag:'1'});
      this.setData({way:way});
    },
    formSubmit:function(e){
      console.log(e);
      var ticket = e.detail.value;
      ticket.way = this.data.way;
      wx.setStorageSync('ticket', ticket);
    }
})
```

从界面提交过来的数据，如图8.37所示。

图8.37 提交表单

这样就完成了弹出框设计，将弹出框选项值赋值给按钮，然后将表单数据提交到后台保存。

8.5 小结

本章主要设计了仿中国婚博会微信小程序，重点掌握以下内容。

（1）学会利用微信小程序来完成界面布局以及给界面添加相关的布局样式。

（2）学会底部标签导航、海报轮播效果、宫格导航的设计。

（3）学会在界面布局的时候，如果有类似或者相同的布局，先设计一个布局和样式，然后再复用这个布局和样式，以提高开发效率。

（4）学会制作下拉菜单筛选条件，动态地切换不同下拉菜单的显示。

（5）学会表单制作，设计表单样式以及表单组件的使用。

（6）学会弹出窗设计，动态地控制显示与隐藏效果。